AF887571

HPE Aruba Networking Certified Associate - Switching
OFFICIAL CERTIFICATION STUDY GUIDE
(EXAM HPE6-A86)

First Edition

Christine Crafton, Henrique Almeida, Sagar Mysore Rajashekhara

HPE Press
660 4th Street, #802
San Francisco, CA 94107

HPE Aruba Networking Switching Associate
Official Certification Study Guide

Christine Crafton, Henrique Almeida, Sagar Mysore Rajashekhara

© 2024 Hewlett Packard Enterprise Development LP

Published by:

Hewlett Packard Enterprise Press
660 4th Street, #802
San Francisco, CA 94107

All rights reserved. No part of this book may be reproduced or transmitted in any form or by any means, electronic or mechanical, including photocopying, recording, or by any information storage and retrieval system, without written permission from the publisher, except for the inclusion of brief quotations in a review.

ISBN: 978-1-960058-24-9

Printed in Mexico

WARNING AND DISCLAIMER

This book provides information about the topics covered in the HPE Aruba Networking Switch-ing Associate certification exam (HPE6-A86). Every effort has been made to make this book as complete and as accurate as possible, but no warranty or fitness is implied.

The information is provided on an "as is" basis. The author, and Hewlett Packard Enterprise Press, shall have neither liability nor responsibility to any person or entity with respect to any loss or damages arising from the information contained in this book or from the use of the discs or programs that may accompany it.

The opinions expressed in this book belong to the author and are not necessarily those of Hewlett Packard Enterprise Press.

FEEDBACK INFORMATION

At HPE Press, our goal is to create in-depth reference books of the best quality and value. Each book is crafted with care and precision, undergoing rigorous development that involves the expert-ise of members from the professional technical community.

Readers' feedback is a continuation of the process. If you have any comments regarding how we could improve the quality of this book, or otherwise alter it to better suit your needs, you can contact us through email at hpepress@epac.com. Please make sure to include the book title and ISBN in your message.

We appreciate your feedback.

Publisher: Hewlett Packard Enterprise Press

HPE Press Program Manager: Michael Bishop

About the authors

Christine Crafton is an instructional designer with over 15 years of experience creating technical training. She is based out of the United States and has been with HPE since 2022.

Henrique Almeida is a Certified HPE Aruba Switching Expert with over 18 years of networking experience. He is based in Brazil and works as an instructional designer and subject matter expert for HPE Aruba Networking Education Services.

Sagar Mysore Rajashekhara is a learning and development specialist with over 10 years of experience in creating end-to-end learning solutions. Sagar is based out of India and works as an instructional designer for HPE.

Introduction

Our dedicated team of instructors—experts in the field—is committed to providing you with hands-on experiences, lab exercises, and in-depth insights into the intricacies of network setup using switches. You will learn about AOS-CX switch technologies including Virtual Local Area Networks (VLANs), secure access using features like redundancy technologies such as Multiple Spanning Tree Protocol (MSTP), link aggregation techniques including Link Aggregation Control Protocol (LACP), and switch virtualization. You will also learn about IP routing, including static and dynamic IP routing with Open Shortest Path First (OSPF).

This guide teaches you the fundamental skills necessary to configure and manage modern, open standards-based networking solutions using HPE Aruba Networking CX Switch Operating System (AOS-CX) routing and switching technologies. It also helps you prepare for the HPE Aruba Networking Switching Associate certification exam (HPE6-A86).

In this guide, we will delve into the principles of network theory and equip you with the skills needed to proficiently configure and manage your network infrastructure. You will learn about AOS-CX switch technologies including Virtual Local Area Networks (VLANs), secure access using features like redundancy technologies such as Multiple Spanning Tree Protocol (MSTP), link aggregation techniques including Link Aggregation Control Protocol (LACP), and switch virtualization. You will also learn about IP routing, including static and dynamic IP routing with Open Shortest Path First (OSPF).

Note: To help assess your understanding of any topic, learning check questions are provided throughout. Refer to the Appendix to verify the correct answers.

Certification and learning

Hewlett Packard Enterprise Certification and Learning provides end-to-end continuous learning programs and professional certifications that can help you open doors and accelerate your career.

We provide

- **Professional sales and technical training and certifications** to give you the critical skills needed to design, manage and implement the most sought-after IT disciplines; and
- **Continuous learning activities and job-role based learning plans** to help you keep pace with the demands of the dynamic, fast paced IT industry
- **Advanced training** to help you navigate and seize opportunities within the top IT transformation areas that enable business advantage today.

As a Certification and Learning certified member, your skills, knowledge, and real-world experience are recognized and valued in the marketplace. To continue your professional and career growth, you have access to our large HPE community of world-class IT professionals, trend-makers and decision-makers. Share ideas, best practices, business insights, and challenges as you gain professional connections globally.

To learn more about HPE Certification and Learning certifications along with continuous learning programs, please visit http://certification-learning.hpe.com.

Audience

This book is designed for people starting a career as networking IT professionals. They support, or operate in a limited scope, wired networks, understanding wired technologies used in SMB, edge, and simple core environments.

Assumed knowledge

It is recommended that candidates have foundational networking experience or attend *Introduction to HPE Aruba Networking Solutions* to obtain a basic knowledge of networking.

Minimum qualifications

Note that there are no prerequisites for taking the HPE6-A86 exam. However, we recommend having already achieved the Aruba Certified Network Technician exam (HPE3-U01).

Relevant certifications

After you pass this exam, your achievement may be applicable toward more than one certification. To determine which certifications can be credited with this achievement, log in to The Learning Center and view the certifications listed on the exam's More Details tab. You might be on your way to achieving additional certifications.

Preparing for exam HPE6-A86

This self-study guide does not guarantee that you will have all the knowledge you need to pass the exam. It is expected that you will also draw on real-world experience and would benefit from completing the hands-on lab activities provided in the associated instructor-led training or self-paced eLearning found at https://inter.viewcentral.com/events/cust/default.aspx?cid=aruba.

Recommended HPE training

Recommended training to prepare for each exam is accessible from the exam's page in The Learning Center. See the exam attachment, "Supporting courses," to view and register for the courses.

Obtain hands-on experience

You are not required to take the recommended courses, and completion of training does not guarantee that you will pass the exam. Hewlett Packard Enterprise strongly recommends a combination of training, thorough review of courseware and additional study references, and sufficient on-the-job experience prior to taking an exam.

Exam registration

To register for an exam, go to https://certification-learning.hpe.com/tr/learn_more_about_exams.html.

CONTENTS

1 AOS-CX switching portfolio

Welcome to the AOS-CX switching portfolio chapter, where we will explore the foundational concepts of networking, powered by the innovative HPE Aruba Networking CX Switch Operating System platform. We will delve into the essentials of networking and the HPE Aruba Networking switching portfolio, designed to equip you with the knowledge needed to navigate and optimize modern network infrastructures.

OBJECTIVES

After completing this chapter, you should be able to:

- ✓ Explain network hierarchical models.
- ✓ Define the attributes of access, aggregation, and core switches.
- ✓ List the major differences in AOS-CX switches.
- ✓ Describe the benefits of the AOS-CX operating system.

Network design

In the realm of network design, architects often rely on hierarchical structures to organize and optimize data flow. This section explores fundamental hierarchies, and by understanding the distinct advantages and considerations of each hierarchy, network designers can strategically plan and implement scalable and resilient network infrastructures tailored to the specific needs of the organization.

Hierarchical model

What is a hierarchical model?

A hierarchical model provides a modular view of the network, simplifies the design, and provides a scalable infrastructure. It refers to a structured approach to organizing network devices and resources into layers or tiers based on their functions and responsibilities. In a hierarchical model, each layer serves a specific purpose and has defined interactions with adjacent layers, leading to a more modular, efficient, and scalable network design.

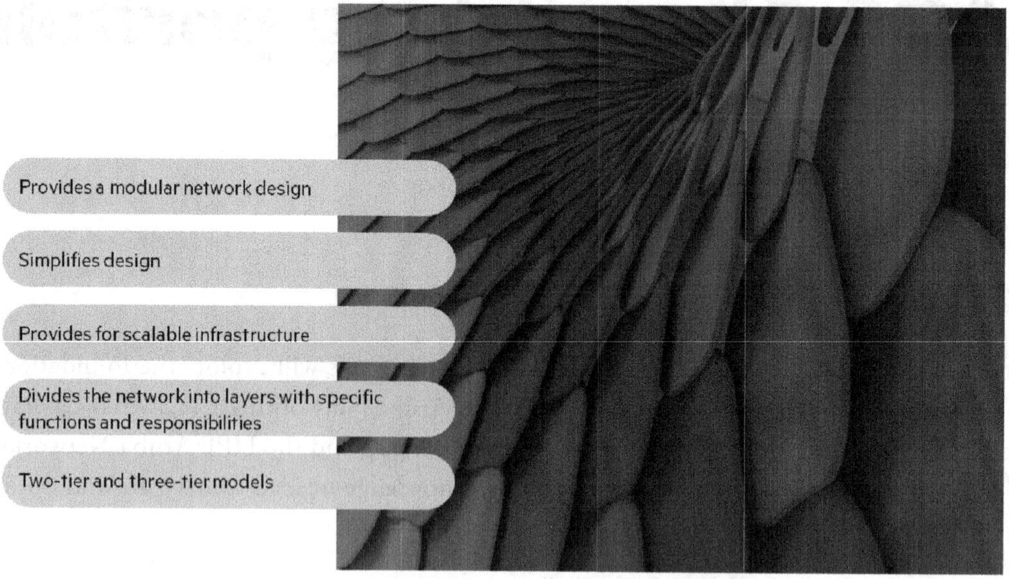

Figure 1-1: Description of hierarchical model

There can be up to three layers in a network hierarchical model:

1. **Access layer:** The access layer, also known as the edge layer, is responsible for connecting access points and end-user devices to the network. It implements virtual LANS (VLANs) to segment traffic, enforces port security policies, applies quality of service (QoS) to prioritize traffic, and may provide Power over Ethernet (PoE) to devices such as wireless access points, surveillance cameras, and IP phones. The access layer also supports link aggregation to combine multiple physical links into a single logical link.
2. **Aggregation layer:** The aggregation layer, also known as the distribution layer, aggregates traffic from multiple access layer devices and provides connectivity to the core layer. It facilitates communication between different network segments and ensuring efficient data flow throughout the network. It provides routing, filtering, and policy-based decisions. The aggregation layer often performs functions such as routing, VLAN configuration, access control and Quality of Services (QoS) enforcement, while also providing redundancy and high availability through link aggregation and redundancy protocols.
3. **Core layer:** The core layer is the backbone of the network, responsible for high-speed data transport between distribution layer devices and connecting different parts of the network. Its primary function is to provide fast and reliable packet forwarding with minimal latency.

Three-tier hierarchy

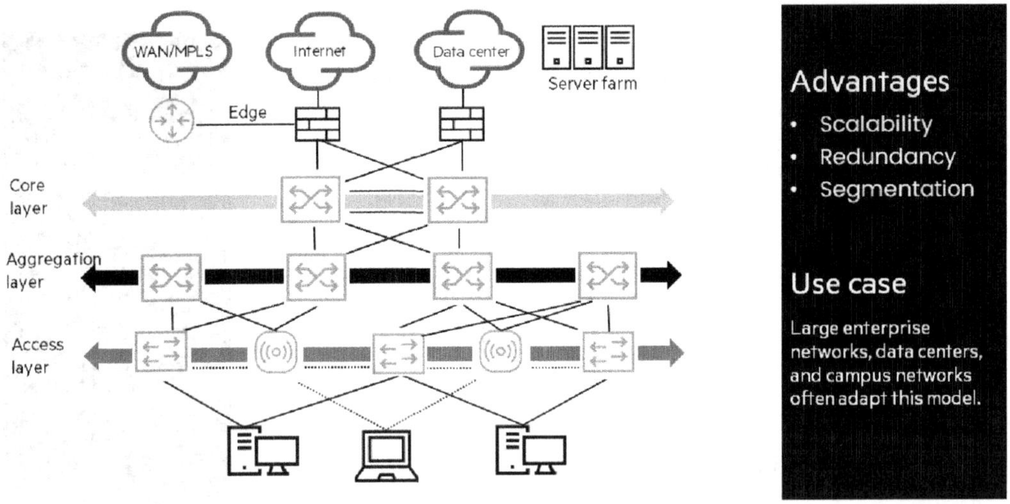

Figure 1-2: Three-tier hierarchy topology

A three-tier design divides the network into three layers—core, aggregation, and access.

The **core layer** is as described for a two-tier design. It must switch and route packets as fast as possible, maintain high availability, and quickly respond to changes in the network. The aggregation layer, edge network, and server farms may connect to this layer.

The **aggregation layer** provides Layer 3 features for traffic to and from the access layer. Routing protocols and ACLs ensure that the correct traffic is routed along the best path. Essentially, you have expanded the core into two separate layers. This can increase cost, and perhaps complexity. However, it may make the network more scalable in certain larger deployments.

The **access layer** provides for endpoint access, with PoE and Layer 2 features, as with a two-tier design.

Two-tier hierarchy

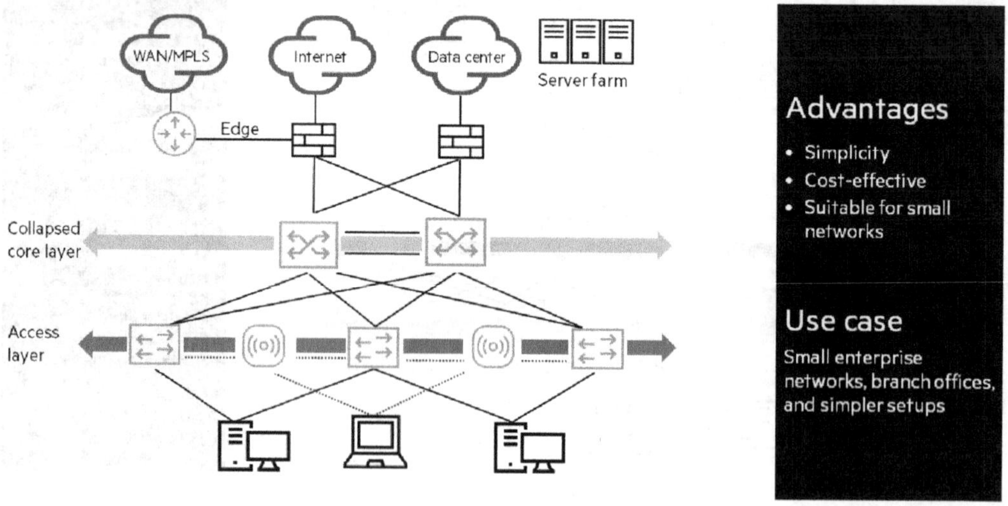

Figure 1-3: Two-tier hierarchy topology

A two-tier design divides the network into two layers—collapsed core and access. Collapsed core is a new term, and what it means in network design, is that it is an architecture where the core and distribution layers of the network are collapsed into a single layer. In this design, access switches connect directly to a collapsed care switch or switches, eliminating the separate distribution layer. The Layer 2 and Layer 3 protocols run close to the endpoints. This design can secure the network by deploying Access Control Lists (ACLs). Quality of Service (QoS) can analyze and prioritize traffic. This ensures the best performance for the most business-critical and delay-sensitive applications.

Wireless networks can also benefit from this two-tier design. Multilayer access switches receive and process critical wireless traffic from Access Points (APs). Appropriate security and QoS policies are applied, and the traffic is routed toward its destination. Due to the separation of the Core and Access layer, this allows for easier scalability, as you can add more APs and expand the network without significantly impacting the core components.

You want three key things at the **core layer – speed, speed, and speed**! This core must switch and route packets as fast as possible. This is why most of the more processing-intensive end-user services are off-loaded to the access layer. You also need reliability at the core. Because of this, high availability (HA) is critical, as is the ability to respond to changes in the network quickly.

Access layer switches have redundant connections to the core to **improve resiliency and mitigate outages**. The **access layer** provides endpoint device connectivity, with Power over Ethernet (PoE), and several Layer 2 and Layer 3 features.

Other sections of the network may also connect to the core. For example, the edge network contains network devices that provide secure connectivity to WAN links and the internet. The server farm contains racks of physical and/or virtual servers and their associated switches and hypervisors.

Campus versus data center network

Figure 1-4: Five differences between campus and data center switches

When it comes to networks, there are campus and data center networks, and different switches are appropriate. Campus switches and data center switches serve different purposes and are optimized for distinct environments, leading to several key differences:

Deployment environments

- Campus switches are typically deployed in enterprise or campus networks, where they connect access points, cameras and end-user devices such as printers, and IP phones to the network infrastructure.
- Data center switches, on the other hand, are deployed in data center environments, where they facilitate the high-speed exchange of data between servers, storage systems, and other network resources.

Traffic patterns

- Campus switches primarily handle client-to-server and client-to-client traffic within the Local Area Network (LAN). They focus on providing connectivity, scalability, and access control for end-user devices.
- Data center switches manage large volumes of east-west traffic, primarily between servers and storage systems within the data center. They prioritize high-speed, low-latency communication to support mission-critical applications and services.

Scale and performance

- Campus switches typically have lower port densities and less emphasis on raw performance, focusing instead on features like PoE, wireless connectivity, and access control for end-user devices.
- Data center switches are designed for high-performance, high-density environments, with features such as high-speed ports, low-latency switching fabrics, and extensive buffering capabilities to accommodate the demanding requirements of data center applications.

Features and functionality

- Campus switches focus on user experience and security. It often include features such as VLAN support, QoS, and ACLs to manage network traffic and enforce security policies at the edge of the network.
- Data center switches focus on high performance and low latency. They prioritize features like Layer 2 and Layer 3 routing, virtualization support, and network virtualization overlays to optimize resource utilization, improve scalability, and streamline network management within the data center.

Redundancy and resilience

- Both campus and data center switches may incorporate redundancy features such as link aggregation (for example, EtherChannel) to ensure high availability and fault tolerance with their respective environments.

Overall, while there may be some overlap in functionality, campus switches and data center switches are designed to address the unique requirements and challenges of their intended deployment environments, leading to differences in design, performance, and feature set.

Network design summary

The three-tier design maximizes flexibility, scalability, stability, and redundancy, while the two-tier design simplifies the network for smaller environments

Figure 1-5: Network design summary

In summary, the three-tier design maximizes flexibility, scalability, stability, and redundancy, while the two-tier design simplifies the network for smaller environments.

Some engineers may choose a two-tier hierarchy, while others might choose a three-tier hierarchy. This decision is based on various factors:

- Number of wired and wireless users
- Mobility requirements
- Number of access points deployed
- Environmental list factors
- Cabling/fibers
- Cost
- Throughput / bandwidth
- Number of buildings on the campus
- Desired security level
- Features and protocols needed

Switching portfolio

It is important to know how the HPE Aruba Networking switching portfolio fits into a network design. The role of the access, aggregation, and core layer in a two- and three-tier hierarchy directly corresponds to the switches placed in those layers. the HPE Aruba Networking switching portfolio has access, aggregation, and core switches. Those switches also have operating systems: AOS-S and AOS-CX. We will briefly cover the AOS-S switches in our switching portfolio, and the remainder of this guide will focus on the AOS-CX switches.

Figure 1-6: HPE Aruba Networking switches grouped together

Legacy network challenges

Traditionally, the network switch had one job: to provide wired access to computing devices. These legacy switches had low port bandwidth utilization and minimal power requirements. Each wired user connected to a separate switch port and each port handled traffic from a single user.

Operational complexity: Dealing with disparate network operating systems, oversubscribed hardware, and complex software licensing overburdens resource-constrained IT teams. The increasing number of Internet of Things (IoT) devices connecting to the network brings new security concerns, which raises the importance for better visibility into what's running on the network and an easier way to segment traffic.

Limited control and visibility: Quickly determining the root cause of an application or network slowdown is challenging. Existing network analytic tools, typically performed by external devices running separately purchased software, provide fragmented data with limited actionable insights. High network traffic volumes and mission-critical network access put growing pressure on IT teams to diagnosis and resolve issues instantly.

Legacy networks cannot keep pace: The shift to digital has disrupted every portion of the network from the edge to the core. As mobile users demand high-performance video, voice, and cloud applications to better collaborate, conduct business, and learn, the amount of data crossing enterprise networks is increasing exponentially. Aging networks are inhibited by closed system architectures with highly manual, hard-coded configs that restrict the adoption of new technologies needed to support time-sensitive networking services and 24x7 access.

Modern switching requirements

Today, the role of the switch has dramatically changed. Most users prefer a wireless connection. The access switch has become a wireless aggregation device. They connect to APs, which in turn connect to switches. Thus, the switch has become an aggregation device, where each port handles traffic from multiple endpoints.

Virtual Extensible LAN (VXLAN) is a modern switching requirement for both data center networks and campus networks. VXLAN enables network virtualization by creating virtual Layer 2 networks (overlay networks) over Layer 3 infrastructure. It extends the scalability of traditional VLANs facilitating flexible deployment of virtual machines and applications in cloud computing environments.

For most organizations, good wireless communication is mission critical. Resilient, reliable, and high-performance aggregation must ensure the best mobile user experience. Also, there is an increasing number of IoT devices—often deployed throughout the organization. To support these devices, switches must offer security, authentication, and services such as PoE to devices that request them.

The PoE feature enables a switch to provide device power, in addition to data connectivity. This eliminates the need to run separate power cables to every device and drastically reduces the cost. However, this means that the switch must have big enough power supplies to not only power itself, but all connected PoE devices.

HPE has created a switching portfolio that meets today's requirements in terms of performance, security, manageability, automation, and wireless optimization. The portfolio accommodates any organization, regardless of its size and complexity.

To meet the objectives of modern network deployments, HPE Aruba Networking has developed two Network Operating Systems (NOS): AOS-S and AOS-CX.

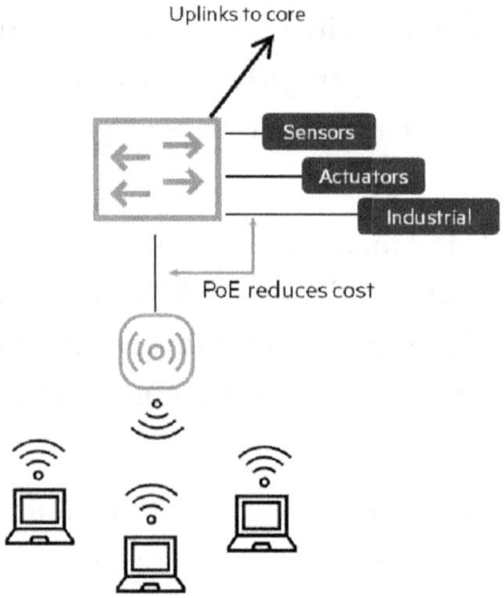

Figure 1-7: Topology of modern switching requirements for wireless connectivity

AOS-S

For HPE Aruba Networking, all the AOS-S are primarily access switches for enterprise. This family of switches offer excellent performance in campus LANs, offering Layer 2 and Layer 3 features.

Most of the installed base from the past is the AOS-S switch. These switches have a ProVision ASIC, and because so many still use these switches, it is important to know how to integrate with them.

All the AOS-S switches offer PoE and support HPE Aruba Networking (Legacy) Management Software, HPE Aruba Networking Central, Web GUI, and CLI. They all utilize REST APIs for automation and analytics.

The biggest differences between the switches are:

- Fixed port and chassis form factors
- Supported stacking technologies
- Speeds and performance

AOS-CX

AOS-CX is a modern, database-driven network switch operating system that delivers automation, distributed analytics, security, and high availability to campus, branch, and data center networks.

Features built for enterprise networks

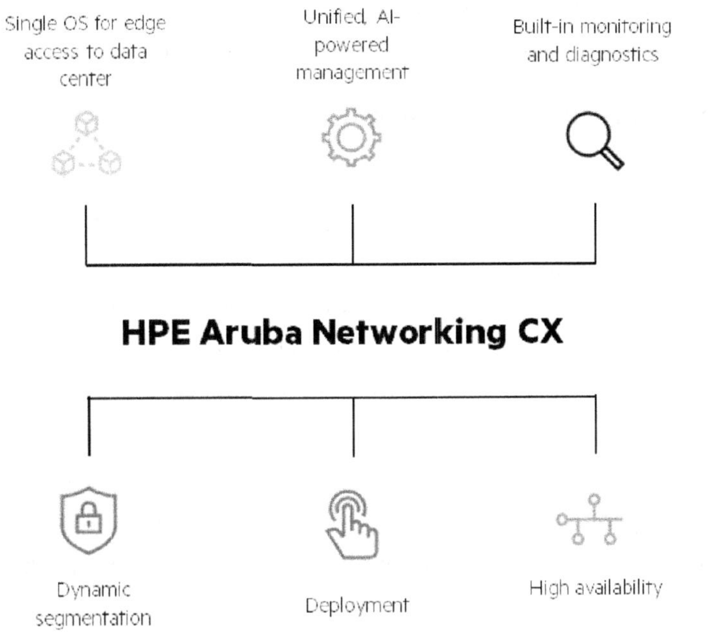

Figure 1-8: Main features for the AOS-CX for enterprise networks

Single OS for edge access to data center

AOS-CX is a modern, database-driven network switch operating system that delivers automation, distributed analytics, security, and high availability to campus, branch, and data center networks. AOS-CX is built with a microservices architecture to automate and simplify IT operations, and it scales from entry level access to high availability data center switching. By eliminating silos of separate switch architectures for different deployments such as branch office and campus, AOS-CX helps eliminate complexity with a consistent operator experience, simplified network design, and unified management across the entire network.

Unified, AI-powered management

HPE Aruba Networking Central is an AI-powered solution that simplifies IT operations, improves agility, and reduces costs by unifying management of all network infrastructures. Built for enterprise-grade resiliency and security, while simple enough for smaller businesses with limited IT staff, it is your single point of visibility and control that spans the entire network—from branch to data center, wired and wireless LAN to WAN. Available as a cloud-based or on-premises solution, HPE Aruba Networking Central is designed to simplify day zero through day two operations with streamlined workflows for tasks such as virtual switch stack creation, automated monitoring using AI-powered insights and NAE, as well as a unified view of all devices and users, both wired and wireless. Comprehensive switch management capabilities include configuration, onboarding, monitoring, troubleshooting, and reporting.

Built-in monitoring and diagnostics

HPE Aruba Networking Network Analytics Engine provides real-time, network-wide insights to swiftly detect, prioritize, and troubleshoot issues as they occur. Rules-based, real-time monitoring and intelligent notifications automatically correlate to configuration changes to help accelerate diagnostic routines and resolve issues before users and business are impacted. Using advanced telemetry and automation built into select AOS-CX switches, network operators can quickly identify and troubleshoot network, system, application, and security-related issues, using Python and CLI-based agents and REST APIs.

A built-in Time Series Database (TSDB) stores configuration and operational state data, making it available to network operators to quickly triage and resolve problems. This data may also be used to analyze historical trends, identify anomalies, and predict future problems due to scale, security, and performance bottlenecks.

HPE Aruba Networking Central uses NAE and agents to deliver switch monitoring, analytics, and enhanced troubleshooting for wired assurance. HPE Aruba Networking Switch Multi-Edit Software and third-party tools such as ServiceNow and Slack provide the intelligence to integrate NAE alerts into IT service management processes, speeding problem resolution.

Dynamic Segmentation

You can secure your network with Dynamic Segmentation, which means having identity-based access privileges to applications and data based on role-based privileges and associated access permissions. This is a fundamental concept of both Zero Trust Security and SASE frameworks.

Also, HPE Aruba Networking Central NetConductor offers cloud-native network security services that enable global policy management and automated network configuration with business-intent workflows. It uses an intelligent EVPN/VXLAN overlay to facilitate distributed Dynamic Segmentation at global scale with policies that are enforced inline and continuously monitored.

Dynamic Segmentation offers a choice of two enforcement models—centralized and distributed—that can coexist and be flexibly adopted. Centralized Dynamic Segmentation uses GRE tunnel-based overlays and is enabled by the policy enforcement firewall that runs natively on the HPE Aruba Networking infrastructure, along with HPE Aruba Networking ClearPass Policy Manager Platform.

You can go beyond traditional identification and profiling techniques with HPE Aruba Networking Central Client Insights. Client Insights is an agentless solution that uses native infrastructure telemetry and machine learning to detect and profile every connected client so you can assign appropriate policies.

Deployment

There are different types of deployment that make the AOS-CX switch easy to use for networking. One is One-touch deployment, and the other is Zero-touch provisioning.

One-touch deployment with the HPE Aruba Networking CX Mobile App means installing switches into a campus network just got a whole lot easier. Use your iOS or Android mobile device to configure, view, and manage AOS-CX switches running 10.04 or newer firmware.

Zero-touch provisioning (ZTP) is a network deployment method that automates the configuration and provisioning of network devices without requiring manual intervention. It allows for seamless and rapid deployment of new devices by automatically detecting and configuring network settings based on predefined templates or policices.

High availability

High availability is a must-have requirement at the core and aggregation layers of campus networks, as well as within leaf-spine architectures in data centers. Requirements for 24x7 network availability leave no windows for upgrades and important configuration changes. HPE Aruba Networking Switching Extension (VSX) has been designed from the ground up to deliver the availability, virtualization, and simplicity requirements for a non-stop, carrier-class network. With unique control plane synchronization for multi-chassis high availability and an architecture that's redundant in both hardware and software, VSX Live Upgrade offers a better way to ensure business success with a network that is always available, even during software upgrades.

AOS-CX switching portfolio

As we talk about AOS-CX models, it is important to remember that the operating system is the same. If you can configure a CX 6300 Series, you will be familiar with the CX 83xx Series or CX 8400 Series. The configuration syntax is shared across the portfolio, making for adding products and different AOS-CX switches easier.

Some of the AOS-CX switches can be configured as a campus switch, while others are focused on data centers. Again, the operating system is the same, but switches vary in features like VXLAN support, speed, performance, and other settings.

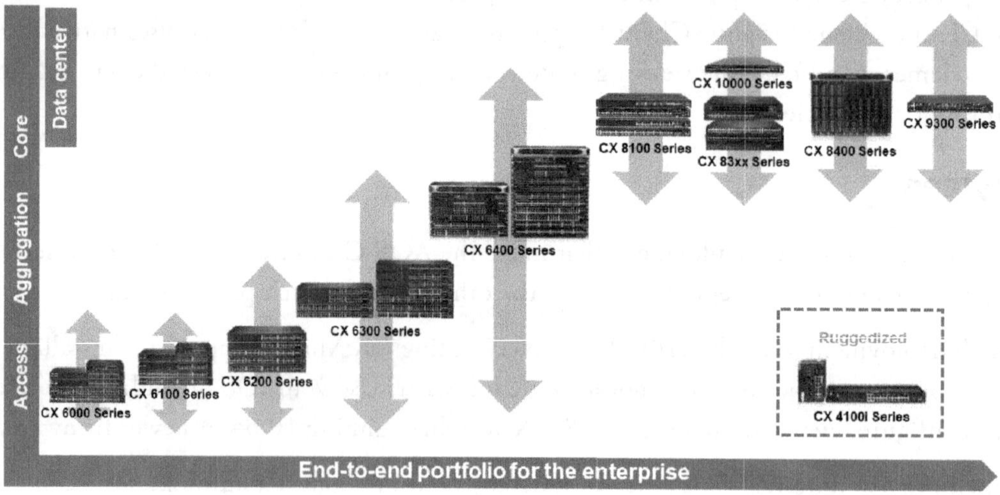

Figure 1-9: The switches running AOS-CX

The following tables show some of the more popular features for each switch. For more detailed information, see the data sheet for each switch.

Table 1-1: CX 4100i Series

	4100i series
Speeds	Down: 1G/10G Up: 1G/10G
Form factor	
Deployment	Ruggedized, access IoT, Harsh environments
Onboard analytics (NAE)[1]	
API and GUI orchestration[2]	✓
REST APIs, Python scripting	✓
Stacking (VSF), Virtualization (VSX)	
Virtualization (VSX/VSF[3]), live upgrade	
Carrier-class HA, deep buffers, large tables	
Dynamic Segmentation	UBT/LUR
Always on PoE	✓
Hot swap, redundant power	

	4100i series
Smart Rate: Multi-gigabit Ethernet	
E-W firewall, microsegmentation	

[1]HPE Aruba Networking Network Analytics Engine

[2] Advanced Network Management with HPE Aruba Networking Central

[3]HPE Aruba Networking Virtual Switching eXtension/HPE Aruba Networking Virtual Switching Framework

Table 1-2: CX 6000 Series, CX 6100 Series and CX 6200 Series

	CX 6000 Series	CX 6100 Series	CX 6200 Series
Speeds	Down: 1G, 1G/10G Up: 1G, 1G/10G	Down: 1G/10G Up: 1G/10G	Down: 1G Up: 1G
Form factor	Compact 1U models	Compact 1U models	Compact 1U models with stacking (8)
Deployment	Access	Access	Access
Onboard analytics (NAE)[1]			✓
API and GUI orchestration[2]	✓	✓	✓
REST APIs, Python scripting	✓	✓	✓
Stacking (VSF), Virtualization (VSX)			VSF (8)
Virtualization (VSX/VSF[3]), live upgrade	✓	✓	
Carrier-class HA, deep buffers, large tables			
Dynamic Segmentation	UBT/LUR		
Always on PoE			✓
Hot swap, redundant power			✓
Smart Rate: Multi-gigabit Ethernet			✓
E-W firewall, microsegmentation			

[1]HPE Aruba Networking Network Analytics Engine

[2]HPE Aruba Networking Fabric Composer

[3]HPE Aruba Networking Virtual Switching eXtension/HPE Aruba Networking Virtual Switching Framework

Table 1-3: CX 6300 Series, CX 6400 Series, and CX 8100 Series

	CX 6300 Series	CX 6400 Series	CX 8100 Series
Speeds	Down: 1G, 2.5G, 5G, 10G Up: 10G, 25G, 40G, 50G, 100G	1G, 10G, 25G, 40G, 100G	10G, 10GBASE-T, and Smart Rate support
Form factor	Compact 1U models with stacking (10)	Efficient 5 or 10-slot (7-12U) modular chassis	Compact 1U models
Deployment	Aggregation, Access Data Center: ToR (1 GbE)	Core, Aggregation, Access Data Center: EoR/MoR	Collapse Core, Small Fabric and ToR/DC POD
Onboard analytics (NAE)[1]	✓	✓	✓
API and GUI orchestration[2]	✓	✓	✓
REST APIs, Python scripting	✓	✓	✓
Stacking (VSF), Virtualization (VSX)	VSF (10)	VSX	VSX
Virtualization (VSX/VSF[3]), live upgrade	✓	✓	✓
Carrier-class HA, deep buffers, large tables			
Dynamic Segmentation			
Always on PoE	✓	✓	
Hot swap, redundant power	✓	✓	✓
Smart Rate: Multi-gigabit Ethernet	✓	✓	✓
E-W firewall, microsegmentation			

[1]HPE Aruba Networking Network Analytics Engine

[2]HPE Aruba Networking Fabric Composer

[3]HPE Aruba Networking Virtual Switching eXtension/HPE Aruba Networking Virtual Switching Framework

Table 1-4: CX 8320 Series, CX 8325 Series, and CX 8360 Series

	CX 8320 Series	CX 8325 Series	CX 8360 Series
Speeds	10G, 40G	10G, 25G, 40G, 100G	10G, 25G, 40G, 100G, 10GBASE-T
Form factor	Compact 1U model	Compact 1U model	Compact 1U model
Deployment	Aggregation, Small Campus, Core Data Center: Spine/Leaf	Core, Aggregation Data Center: Spine/Leaf	Core, Aggregation Data Center: Spine/Leaf
Onboard analytics (NAE)[1]	✓	✓	✓
API and GUI orchestration[2]	✓	✓	✓
REST APIs, Python scripting	✓	✓	✓
Stacking (VSF), Virtualization (VSX)	VSX	VSX	VSX
Virtualization (VSX/VSF[3]), live upgrade	✓	✓	✓
Carrier-class HA, deep buffers, large tables			
Dynamic Segmentation			
Always on PoE			
Hot swap, redundant power	✓	✓	✓
Smart Rate: Multi-gigabit Ethernet	✓		✓
E-W firewall, microsegmentation			

[1]HPE Aruba Networking Network Analytics Engine

[2]HPE Aruba Networking Fabric Composer

[3]HPE Aruba Networking Virtual Switching eXtension/HPE Aruba Networking Virtual Switching Framework

Table 1-5: CX 8400 Series, CX 9300 Series, and CX 10000 Series

	CX 8400 Series	CX 9300 Series	CX 10000 Series
Speeds	10G, 40G, 100G	100G, 200G, 400G	10G, 25G, 100G
Form factor	Effcient 8-slot (8U) modular chassis	Compact 1U model	Compact 1U model

	CX 8400 Series	CX 9300 Series	CX 10000 Series
Deployment	Core, Aggregation Data Center: EoR/MoR	Core, Aggregation Data Center: Spine/Leaf	Core, Aggregation Data Center: Leaf w/FW
Onboard analytics (NAE)[1]	✓	✓	✓
API and GUI orchestration[2]	✓	✓	✓
REST APIs, Python scripting	✓	✓	✓
Stacking (VSF), Virtualization (VSX)	VSX	VSX	VSX
Virtualization (VSX/VSF[3]), live upgrade	✓	✓	✓
Carrier-class HA, deep buffers, large tables	✓	✓	
Dynamic Segmentation			
Always on PoE			
Hot swap, redundant power	✓	✓	✓
Smart Rate: Multi-gigabit Ethernet			
E-W firewall, microsegmentation			✓

[1]HPE Aruba Networking Network Analytics Engine

[2]HPE Aruba Networking Fabric Composer

[3]HPE Aruba Networking Virtual Switching eXtension/HPE Aruba Networking Virtual Switching Framework

Access layer switch family

	CX 6000	CX 6100	CX 6200	CX 6300M
Models	1U: 12. 24. 48 port	1U: 12. 24. 48 port	1U: 12. 24. 48 port fixed & modular switch models (PSU. Fan Trays)	1U: 12. 24. 48 port fixed & modular switch models (PSU. Fan Trays0
Switching & Routing Services	Layer 2 with static routing	Layer 2 with static routing	Layer 3	Advanced Layer 2 and 3
Switching & Forwarding Performance	Up to 104 Gbps Up to 77.3 Mpps	Up to 176 Gbps	Up to 272 Gbps Up to 202 Mpps	Up to 1760 Gbps Up to 1310 Mpps
Downlink & Uplink Speeds	Down: 1G Up: 1G	Down: 1G/10G Up: 1G/10G	Down: 1G. 1G/10G Up: 1G. 1G/10G	Down: 1G/2.5G/5G/10G Up: 10G/25G/40G/50G/100G
SmartRate Multi-gigabit Support	No	No	✓	✓
Power over Ethernet	Up to Class 4: 30W /Port 740W /Max	Up to Class 4: 30W /Port 740W /Max	Up to Class 6: 30W. 60W /Port 1440W /Max	Up to Class 8: 30W. 60W. 90W /Port 2880W /Max
Always-On PoE	No	No	✓	✓
Dynamic Segmentation	No	No	✓	✓
Stacking & Virtualization	No	No	Stacking Maximum: 8 Switches (VSF)	Stacking Maximum: 10 Switches (VSF)
Automation & Analytics	Rest APIs. Python Scripting. NetEdit	Rest APIs. Python Scripting. NetEdit	Network Analytics Engine (NAE) Rest APIs. Python Scripting. NetEdit	Network Analytics Engine (NAE) Rest APIs. Python Scripting. NetEdit
Management Support	Central. Web GUI. CLI. NetEdit	Central. Web GUI. CLI. NetEdit	Central. Web GUI. CLI. NetEdit	Central. Web GUI. CLI. NetEdit

Figure 1-10: Access layer switch family

Aggregation layer switch family

	CX 6300M	CX 6400
Models	1U: 12, 24, 48 port fixed & modular switch models (PSU, Fan Trays0	Chassis: 5-slot (7 RU) / 10-slot (12 RU) Line Cards: 12, 24, 48 port
Switching & Routing Services	Advanced Layer 2 and 3	Advanced Layer 2 and 3
Switching & Forwarding Performance	Up to 1760 Gbps Up to 1310 Mpps	Up to 28 Tbps Up to 11.4 Bpps
Downlink & Uplink Speeds	Down: 1G/2.5G/5G/10G Up: 10G/25G/40G/50G/100G	1G/10G/25G/40G/100G
SmartRate Multi-gigabit Support	✓	✓
Power over Ethernet	Up to Class 8: 30W, 60W, 90W /Port 2880W /Max	Up to Class 6: 30W, 60W /Port 8760W /Max
Always-On PoE	✓	✓
Dynamic Segmentation	✓	✓
Stacking & Virtualization	Stacking Maximum: 10 Switches (VSF)	Stacking Maximum: 2 switches (VSX)
Automation & Analytics	Network Analytics Engine (NAE) Rest APIs, Python Scripting, NetEdit	Network Analytics Engine (NAE) Rest APIs, Python Scripting, NetEdit
Management Support	Central, Web GUI, CLI, NetEdit	Central, Web GUI, CLI, NetEdit

Figure 1-11: Aggregation layer switch family

If the environment is small enough the CX 6300 Series can be used in an aggregation layer. But CX 6300 Series is mainly used in an access layer.

The CX 6400 Series is mostly used as collapsed core.

Core layer switch family

	CX 6400	CX 8100	CX 83xx	CX 8400	CX 9300	CX 10000
Models	Chassis: 5-slot (7 RU) / 10-slot (12 RU) Line Cards: 12, 24, 48 port	1U, Varies	Varies	8 slot chassis	1U Fixed: 32 ports of 100/200/400GBE	Fixed: 48 port of 10/25GbE 6 port o 10GbE
Switching & Routing Services	Advanced Layer 2 and 3	Switching & routing Limited Layer 2 and 3	Advanced Layer 2 and 3	Advanced Layer 2 and 3	Advanced Layer 2 and 3	Advanced Layer 2 and 3
Switching & Forwarding Performance	Up to 28 Tbps Up to 11.4 Bpps	1.75 Tbps	Varies	19.2 Tbs	25.6 Tbps (FD)	3.6 Tbs
Downlink & Uplink Speeds	1G/10G/25G/40G/100G	1G, 10G,	Varies	10G, 40G, 100G	100G, 200G, 400G	10G, 25G, 100G
SmartRate Multi-gigabit Support	✓	✓	N/A	N/A	N/A	N/A
Power over Ethernet	Up to Class 6: 30W, 60W /Port 8760W /Max	No	No	No	No	No
Always-On PoE	✓	N/A	N/A	N/A	N/A	N/A
Dynamic Segmentation	✓	No	No	No	No	No
Stacking & Virtualization	Stacking Maximum: 2 switches (VSX)	Stack Maximum: 2 switches (VSX)	Stack Maximum: 2 switches (VSX)	Stack Maximum: 2 switches (VSX)	Stack Maximum: 2 switches (VSX)	Stack Maximum: 2 switches (VSX)
Automation & Analytics	Network Analytics Engine (NAE) Rest APIs, Python Scripting, NetEdit	Fabric Composer, Network Analytics Engine (NAE) Rest APIs, NetEdit	Varies	Network Analytics Engine (NAE) Rest APIs., NetEdit	Fabric Composer, Network Analytics Engine (NAE) Rest APIs, NetEdit	Fabric Composer, Network Analytics Engine (NAE) Rest APIs, NetEdit
Management Support	Central, Web GUI, CLI, NetEdit	Web GL Fabric Composer, CLI, NetEdit	Varies	AirWave, Web GUI, CLI, IMC, NetEdit	Web GUI, Fabric Composer, CLI, NetEdit	Web GUI, Fabric Composer, CLI, NetEdit

Figure 1-12: Core layer switch family

HPE Aruba Networking switch ASICs

AOS-CX switches utilize the Aruba Network Processor (ANP) ASIC. The ANP ASIC is specifically designed to enhance the performance and capabilities of AOS-CX switches. Here are some advantages of using the ANP ASIC in these switches:

High Performance: The ANP ASIC is engineered to deliver high throughput and low latency, enabling the switches to handle large volumes of data efficiently. This is particularly important in environments with heavy traffic loads and high-speed networking requirements.

Programmability: The ANP ASIC supports programmability, allowing network operators to implement custom features and policies directly in the hardware. This flexibility enables rapid adaptation to changing network demands and the deployment of new services without requiring hardware changes.

Advanced Telemetry: One of the standout features of the ANP ASIC is its capability for advanced telemetry. This allows for real-time monitoring and analytics of network traffic, providing valuable insights into network performance, potential issues, and areas for optimization.

Enhanced Security: The ANP ASIC includes built-in security features that help protect the network from various threats. These features can include hardware-based encryption, access control mechanisms, and anomaly detection, contributing to a more secure network environment.

Network Automation: The programmability and telemetry features of the ANP ASIC facilitate network automation. By providing detailed data and enabling programmable actions, the ASIC supports automated network management and optimization, reducing manual intervention and the potential for human error.

Scalability: The ANP ASIC is designed to scale with the needs of the network. Whether for small enterprise deployments or large data center environments, the ASIC can handle increasing traffic loads and more complex networking requirements without compromising performance.

Energy Efficiency: Similar to other ASICs, the ANP ASIC is optimized for specific tasks, which leads to better energy efficiency. This helps reduce the overall power consumption of the network infrastructure, contributing to lower operational costs.

Integrated Features: The ANP ASIC integrates various networking functions into the hardware, such as Quality of Service (QoS), traffic management, and routing. This integration ensures these functions are performed at high speed and with minimal latency, improving overall network performance.

Consistency and Reliability: With the specialized design of the ANP ASIC, AOS-CX switches offer consistent and reliable performance. This is crucial for maintaining stable network operations, especially in mission-critical environments.

Future-Proofing: The programmability and flexibility of the ANP ASIC ensure that AOS-CX switches can adapt to future networking standards and technologies. This helps protect the investment in network infrastructure by ensuring it remains relevant and capable of supporting new requirements.

Overall, the ANP ASIC provides AOS-CX switches with a combination of high performance, flexibility, advanced features, and efficiency, making them well-suited for modern, dynamic networking environments.

Switch product label

The switch product label is an orange-colored tab on the bottom right side of the switch front panel. Pull the tab out to view the product label information.

The product label information includes the part number, serial number, and MAC address. Serial numbers and MAC address labels are duplicated on the back or bottom of the product.

Being able to identify this information is important for two main purposes:

- Inventory control
- To open a case with HPE Networking Support for support/warranty

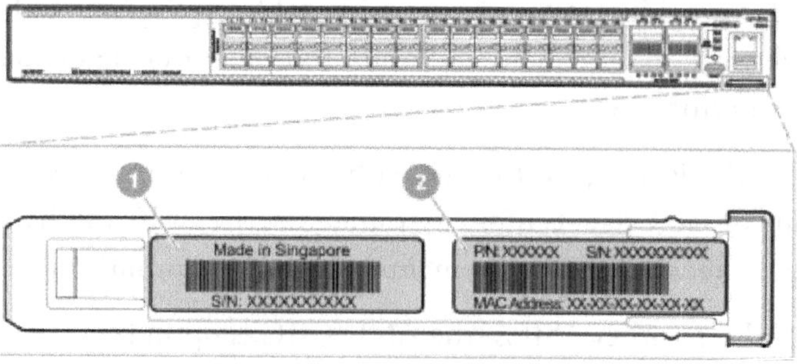

Table 1-6: Switch product label

Label	Description
1	Country of origin and serial number
2	Part number, serial number, and MAC address

Management, monitoring, and orchestration options

Managing a data center network can present some challenges to IT staff, but with HPE Aruba Networking tools, technicians can monitor, manage, configure, and orchestrate the data center network.

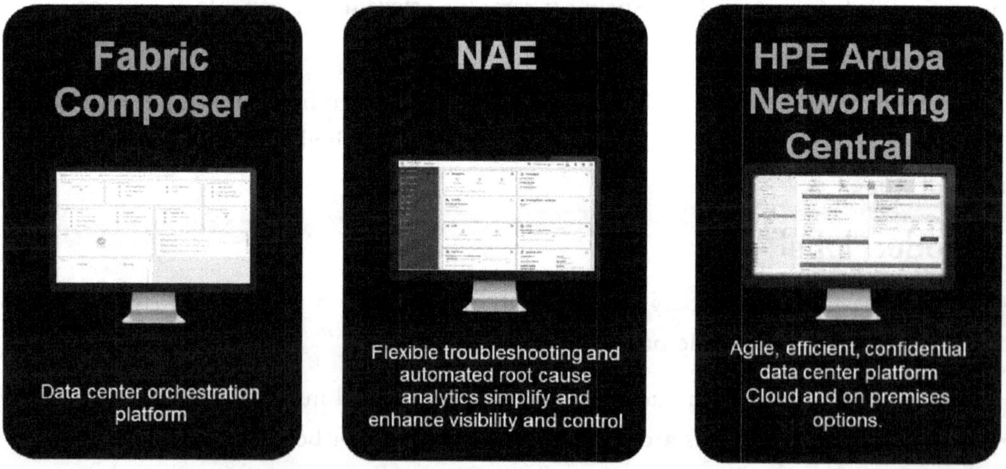

Figure 1-13: Management, monitoring, and orchestration software from HPE Aruba Networking

Fabric Composer: An intelligent, API based, software-defined orchestration solution that simplifies and accelerates leaf-spine network provisioning and day-to-day operations across rack-scale compute and storage infrastructure.

HPE Aruba Networking Network Analytics Engine (NAE): Leverages AOS-CX for intelligent monitoring and automated data collection and analysis. NAE has a wide range of uses—system health, network analytics, security application visibility, and network optimization.

HPE Aruba Networking Central and HPE Aruba Networking Central On-Premises: SaaS platforms that offer you a single intelligent console to monitor, analyze, and configure WLAN and wired networks. HPE Aruba Networking Central makes it easy and efficient to manage your networks by combining industry-leading functionality with an intuitive user interface and enables network administrators and help desk staff to support and control the network for your premises. The features are designed to manage, maintain, and analyze your network. HPE Aruba Networking Central On-Premises is for organizations that do not want the data on cloud because of their business operations and policies regarding confidentiality, security, downtime, loss of data, and so on. This software solution also gives the same agility and efficiency of cloud services as HPE Aruba Networking Central, and at the same time, it adheres to stringent regulatory and compliance requirements.

Learning check

Check your understanding of this material by answering the following questions.

Question 1

What is the primary difference between a data center network and a campus network?

- ○ Campus networks are typically optimized for high bandwidth and low latency, while data center networks are designed to support a large number of users and devices over a broader geographic area.
- ○ Data center networks prioritize wireless connectivity, whereas campus networks prioritize wired connections.
- ○ Data center networks are only used for cloud services, while campus networks are used for local area networks (LANs).
- ○ Data center networks are typically optimized for high bandwidth and low latency, while campus networks are designed to support a large number of users and devices over a broader geographic area.

Question 2

Which CX model allows for VSF?

- ○ CX 10000 Series
- ○ CX 6200M Series, and CX 6300 Series
- ○ CX 6000 Series, CX 6100 Series, and CX 6200F Series

- CX 8100 Series and CX 8400 Series

2 Switch fundamentals

Welcome to the switch fundamentals chapter, where we set up and use our AOS-CX switches. We will learn and utilize AOS-CX CLI code and how it is used for setup, configuration and troubleshooting diagnostic outages. After unboxing the switches, it is time to get the appropriate connections and then all of the various CLI commands required to name, get switch information, conduct basic configurations, discovering the network and more.

OBJECTIVES

After completing this chapter, you should be able to:

- ✓ Discover switch information with CLI `show` commands.
- ✓ Perform basic configuration tasks.
- ✓ Obtain network status and connectivity.
- ✓ Discern switch provisioning methods.

AOS-CX CLI access

The initial access of a switch is crucial for setting up its configuration and accessing its management interface. This includes configuring correct port numbering, familiarizing oneself with various CLI prompts, and understanding how to utilize context-sensitive help, all of which form the foundational steps to effectively using the switch in a network environment.

Management access

When you initially configure a switch, you will typically use the console port or the out-of-band Management (OOBM) interface. A console port on a switch is a physical interface that allows you to directly connect to the switch for configuration and management purposes. It usually looks like a small rectangular socket, often located on the front or back panel of the switch. It is commonly labeled as "Console" or "Console Port" near the port itself.

For either port type, you must establish a serial connection between the management station (perhaps your laptop) and the switch. To do this, you need the following items:

- If the switch has an RJ-45 port, you can use a serial cable (shipped with the switch). A USB-to-Serial converter may be needed since modern laptops no longer include a serial port.
- Software on your laptop or management station that emulates the serial session is also needed. Some common examples include PuTTY, Tera Term, Secure CRT, iTerm, and iTerm2.
- If the switch has a USB-C port, you will use a USB-C cable. Software on your laptop or management station that emulates the serial session is also needed. Some common examples include PuTTY, Tera Term, Secure CRT, iTerm, and iTerm2.

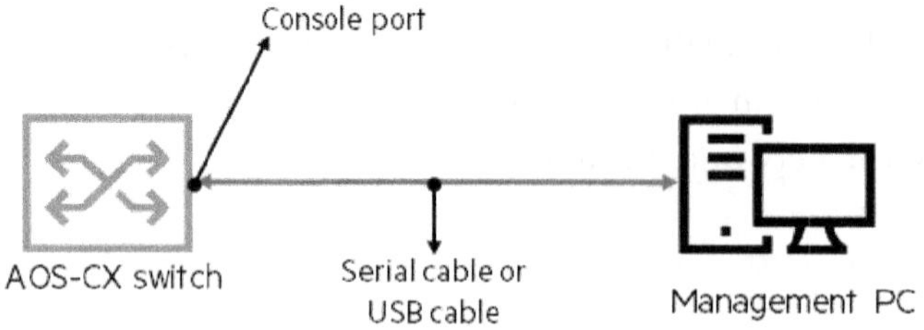

Figure 2-1: Connect switch to console port.

Using either connection, the serial session must be configured with the following parameters (these are the default settings on the switch):

- Baud rate: 115200
- Data bits: 8
- Parity: None
- Stop bits: 1
- No flow control

User groups

The switch provides the following built-in user groups with corresponding roles. Each of these roles comes with a set of privileges.

Administrator

Administrators have full privileges, including:

- Full CLI access.
- Performing firmware upgrades.

- Viewing switch configuration information, including sensitive information such as passwords which are displayed as ciphertext.
- Performing switch configuration.
- Adding/removing user accounts.
- Configuring users accounts, including passwords. Once set, a password cannot be deleted or set to empty.
- REST API: All methods (GET, PUT, POST, DELETE) and switch resources are available.

The privilege level for administrators is 15.

Operators

Operators have no switch configuration privileges. Operators are restricted to:

- Basic display-only CLI access.
- Viewing of nonsensitive switch configuration information.
- REST API: Other than the \login and \logout resources, only the GET method is available.

The privilege level for operators is 1.

Auditors

Auditors are restricted to functions related to auditing only:

- CLI: Access to commands in the auditor context (auditor>) only.
- Web UI: Access to the System > Log page only.
- REST API: POST method available for the \login and \logout resources. GET method available for the following resources only:
    - Audit log: /logs/audit
    - Event log: /logs/event

The privilege level for auditors is 19.

User-defined user groups

The switch enables you to create up to 29 user-defined local user groups, for the purpose of configuring local authorization. Each of the 29 user-defined groups support up to 1024 CLI command authorization rules that define what CLI commands can be executed by members of the group.

The local user groups with their command execution rules are useful for the following:

- Providing authorization for use with RADIUS servers.
- Providing fallback authorization for use with TACACS+ servers.
- Providing authorization when neither RADIUS or TACACS+ servers are used.

Context levels

AOS-CX is organized into different configuration contexts or levels. Each context determines which parts of the switch can be managed and which commands are available to users with the appropriate authority. The operating system defines the following contexts:

Operator context (>)

The operator context enables you to execute commands to view—but not change—configuration. This context requires the least user privilege to execute commands. When in operator context, the CLI prompt is the switch name, followed by a greater than sign (>), as shown in the slide.

Manager context (#)

This is where you can execute commands that do not require saving changes to the configurations. When in manager context, the CLI prompt is the switch name followed by a hashtag.

To navigating to the manager context, start at the operator context (>). Then enter the `enable` command. You must have administrator authority to enter the `enable` command.

```
Switch> enable
Switch#
```

Global configuration context (config)

This is where you can execute commands that change the switch configuration, as shown. To access this mode, start in Manager mode, then enter the `configure terminal` command, or shorten it to just `config`. To move back up one level from this or any other context, issue the `exit` command.

Other configuration command context

All other configuration command contexts are descendants of the global configuration command context (config). From these command contexts, you can execute commands that apply to a specific configuration or protocol such as an interface or a VLAN.

Examples:

```
Switch(config-if)#
Switch(config-router)#
Switch(config-vlan-100)#
```

To return back to the manager context from any child or descendent context, enter the `end` command.

Example:

```
Switch(config-vlan-100)# end
Switch#
```

Table 2-1: Context level descriptions

Prompt	Context	Description
`Switch>`	Operator	View commands, no configuration commands.
`Switch#`	Manager	Use commands that need not be saved.
`Switch(config)#`	Global config	Change switch configuration.
`Switch(config-if)#`	Interface	Descendants of global configuration use commands that apply to a specific configuration or protocol.
`Switch(config-router)#`	Router config	
`Switch(config-vlan-100)#`	VLAN 100 config	

Configuring ports

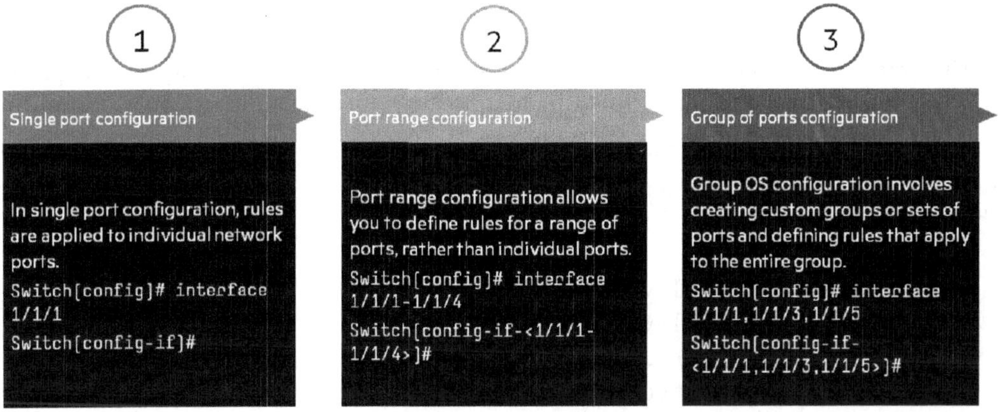

Figure 2-2: Port configuration

Single Port, Port Range and Group of Ports (Group OS) configuration are methods used to define rules or settings for network ports on AOS-CX switches.

Single port configuration: In single port configuration, rules are applied to individual network ports. For example, if you have a port connected to a printer or other device with specific needs, you enter the specific port. But mays you have a group of devices that shares the same needs, you may then use port groups OS port ranges.

Port range configuration: Port range configuration allows you to define rules for a range of ports, rather than individual ports. This is useful when you want to apply the same rule to multiple contiguous ports. For instance, you might configure a firewall to allow traffic on ports 8000 to 9000 for a range of applications or services.

Group of ports configuration (Group OS): Group OS configuration involves creating custom groups or sets of ports and defining rules that apply to the entire group. This allows you to apply rules to multiple non-contiguous ports that serve a common function or are related in some way. For example, you could create a group called "Email Services" that includes ports 25 (SMTP), 110 (POP3) and 143 (IMAP), and then configure rules to allow traffic on all ports within the group for email services.

Context-sensitive help using "?"

The AOS-CX CLI provides you with built-in help features. For example, to show the available commands that you can execute in the current command context, enter the question mark (?) symbol at the CLI prompt. The question mark (?) does not display on the screen when you enter it.

```
Switch# ?
boot                    Reboot all or part of the system
checkpoint              Checkpoint information
<<lines omitted due to space limitation>>
Switch#
```

To show the available parameters for a command, enter the command followed by a space and then enter the question mark symbol (?). This is shown in the bottom example in the figure. Notice that after the CLI displays the information, it automatically displays the text you entered before without including the help symbol (?).

```
Switch(config)# access-list ?
all         All access-lists
ip          Internet Protocol v4 (IPv4)
ipv6        Internet Protocol v6 (IPv6)
log-timer   Set ACL log timer length (frequency)
mac         Ethernet MAC Protocol
Switch(config)# access-list
```

Context abbreviation and completion

AOS-CX supports both command abbreviation and command completion.

Command abbreviation

You need only enter enough letters to uniquely specify a command.

To save time by only typing an abbreviated version of the full syntax, enter enough letters to uniquely specify a valid command, and the CLI accepts the command. For example, you can enter `con` instead of `configure` to navigate from the manager context to the global configuration context.

Command completion

- Enter part of command, press **tab**; CLI completes it.
- Press **tab** twice to see command options.

Command completion means that if you enter part of command word and then press the tab key, one of the following occurs:

- If you enter enough letters to match a valid command, the CLI displays the remainder of the word.
- If you have not entered enough letters to match a valid command, the CLI does not complete the command.

Example:

```
Switch(config)# int + Tab
Switch(config)# interface
```

If you press the tab key twice after a completed word, the CLI displays the command options.

Example:

```
Switch(config)# interface + Tab + Tab
lag        loopback  mgmt      tunnel     vlan       vxlan
```

Basic configuration

Basic switch configuration involves assigning a hostname to the device to uniquely identify it within the network, enabling interfaces to establish network connections, and verifying the status and configurations of these interfaces to ensure proper functionality. This initial setup is essential for establishing network connectivity and laying the foundation for further configuration and management tasks.

Configuring hostname

Providing a name for your hostname is essential as it serves as a unique identifier for your device within a network environment. This name helps in distinguishing your device from others on the network and facilitates easier identification and communication. Additionally, assigning a meaningful hostname can improve network management tasks, such as monitoring, logging, and troubleshooting, by providing context and clarity to network administrators.

Many corporations have documented, standardized naming conventions. For example, one standard might be Building-Floor-Rack_Name. So, the switches in Rack 2, on the 2nd floor of building 9 might be named 9-2-2_SWA, 9-2-2_SWB, 9-2-2_SWC, and so on.

The CLI shows how to name devices, using the `hostname` command.

```
Switch(config)# hostname Access-1
Access-1(config)#
```

Enabling an interface

Enabling an interface via the CLI involves accessing the Configuration mode and issuing commands specific to the interface, followed by the `no shutdown` command to activate the interface. This process enables network connectivity on the interface, allowing it to send and receive data packets within the network environment.

In AOS-CX, all ports on VSX-capable switches are disabled by default. For other switches, the ports are enabled by default.

Enabling interface CLI:

```
Switch(config)# interface 1/1/1
Switch(config-if)# no shutdown
```

Disabling interface CLI:

```
Switch(config)# interface 1/1/1
Switch(config-if)# shutdown
```

Interface description CLI:

```
Switch(config)# interface 1/1/1
Switch(config-if)# description To-Core
```

Verifying an interface status

The `show interface` command shows the status and configuration for all interfaces on the switch.

In this CLI example, "Administratively down" means that a `shutdown` command was configured on that interface.

> Depending on the series, the ports are either enabled or disable by default. Please check your product data sheet.

You can also see that an interface description was configured on that interface with the `description` command.

```
Switch# show interface 1/1/1

Interface 1/1/1 is down (Administratively down)
```

```
Admin state is down
State information: Administrativley down
Link transitions: 0
Description: To-Core
Hardware: Ethernet, MAC Address: 90:20:c2:bc:3f:00
MTU 1500
Type SFP-BT
```

Notice in this example, the interface is now up. The `no shutdown` command has been entered. The interface description remains unmodified.

```
Switch# show interface 1/1/1

Interface 1/1/1 is up
Admin state is up
Link transitions: 0
Description: To-Core
Hardware: Ethernet, MAC Address: 90:20:c2:bc:3f:00
MTU 1500
Type SFP-BT
```

Discovering the network

Discovering the network with Link Layer Discovery Protocol (LLDP) involves enabling LLDP on network devices to exchange information about neighboring devices and their capabilities. By broadcasting and receiving LLDP messages, devices can dynamically discover neighboring network devices, including switches, routers, and endpoints, facilitating network topology discovery and device management. This helps you gain visibility into the network infrastructure, identify connected devices, and troubleshoot connectivity issues more efficiently.

Link layer discovery protocol

Imagine that you are remotely configuring an AOS-CX switch in a production environment. This switch is connected to multiple network devices, and you must configure only the interface that is facing an HPE Aruba Networking Mobility Controller. The onsite IT team is not available, and somehow you must identify the correct interface so you can apply the configuration. This problem is easily addressed using LLDP.

LLDP is an industry standard link layer protocol (Layer 2 protocol). It is used by network devices to advertise their identity and capabilities over a wired Ethernet connection. This protocol enables you to discover and document network device interconnections.

Media Endpoint Discovery is an enhancement of LLDP known as LLDP-MED. This provides for auto-discovery of LAN policies such as VLAN-ID, Layer 2 priority, and Differentiated Services for QoS. This helps to enable plug and play networking.

Devices send LLDP information on a regular interval, encapsulated in Ethernet frames. Directly connected devices receive these frames and store the information in a table in local memory.

You can view this information with commands like `show lldp neighbor-info`.

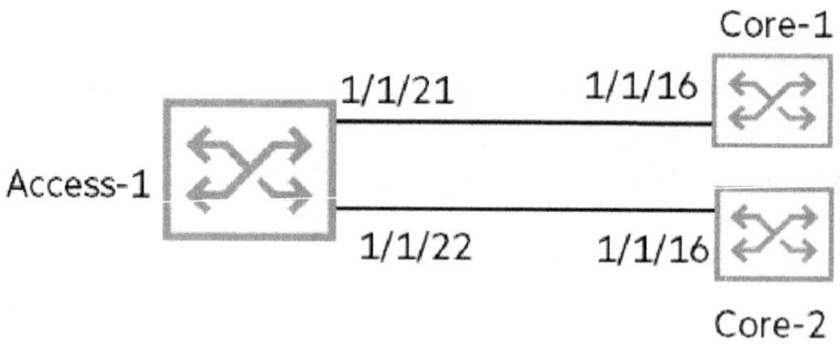

Figure 2-3: LLDP scenario

In the scenario, Access-1's port 1/1/21 connects to port 1/1/16 of Core-1. Access-1's port 1/1/22 connects to port 1/1/16 of Core-2. If someone has already taken the time to create a diagram like the one shown in the figure, you do not need to use LLDP for discovery—it has already been done for you. However, network documentation is often incorrect, out of date, or was simply never created. You can verify existing documentation and create new documentation using LLDP information.

Creating and maintaining good, accurate network documentation is very helpful. When the network fails, instead of having to discover the network, everyone can look at the document and be far more effective troubleshooters. Documentation is an insurance policy against slow, ineffective troubleshooting. It can be the difference between a 4-hour network outage and a 15-minute outage.

In AOS-CX, LLDP is enabled by default.

The `show lldp neighbor-info` command displays information about neighboring devices for all interfaces or for a specific interface.

```
Access-1# show lldp neighbor-info

LLDP Neighbor Information
=========================
<lines omitted due to space limitation>

LOCAL-PORT  CHASSIS-ID         PORT-ID   PORT-DESC      TTL   SYS-NAME
-------------------------------------------------------------------------
1/1/21      90:20:c2:bc:a8:00  1/1/16    1/1/16         120   Core-1
1/1/22      90:20:c2:bb:8b:00  1/1/16    1/1/16         120   Core-2
```

To get more information about a specific entry, you can append the interface at the end of the previous command `show lldp neighbor-info 1/1/21`, as shown below:

```
Access-1# show lldp neighbor-info 1/1/21

Port                            : 1/1/21
Neighbor Entries                : 1
Neighbor Entries Deleted        : 0
Neighbor Entries Dropped        : 0
Neighbor Entries Aged-Out       : 0
Neighbor Chassis-Name           : Core-1
Neighbor Chassis-Description    : Aruba JL635A  GL.10.04.0001
Neighbor Chassis-ID             : 90:20:c2:bc:a8:00
Neighbor Management-Address     : 10.250.1.3
Chassis Capabilities Available  : Bridge, Router
Chassis Capabilities Enabled    : Bridge, Router
Neighbor Port-ID                : 1/1/16
Neighbor Port-Desc              : 1/1/16
Neighbor Port VLAN ID           : 1107
TTL                             : 120
```

LLDP and LLDP-MED

LLDP is an open-standard link layer protocol used by network devices to advertise their identity, capabilities, and neighbors on a local area network. Based on IEEE 802 technology, you use it verify layer-2 connectivity between devices, discover an unknown network topology, and help build network documentation.

The MED is an enhancement of LLDP, known as LLDP-MED. It helps to:

- Provision devices.
- Determine correct power settings for connected VoIP phones and APs.
- Auto-assign VLAN and QoS setting.
- eEnable plug-and-play networking.

Features include:

- Define LAN policies based on the connected device: VLAN, layer 2 802.1p class of service (CoS) values, layer 3 differential service code points (DSCP) settings, and more.
- Discover the location of the device for VoIP and enhanced 911 services.
- Extended and automated power management of power-over-Ethernet (PoE) edge devices.

- Inventory management allows you to track your network devices, and determine device-specific information, like the manufacturer, software and hardware versions, serial or asset number.
- Zero-touch provisioning (ZTP) provides support for quickly deploying new devices in a network.

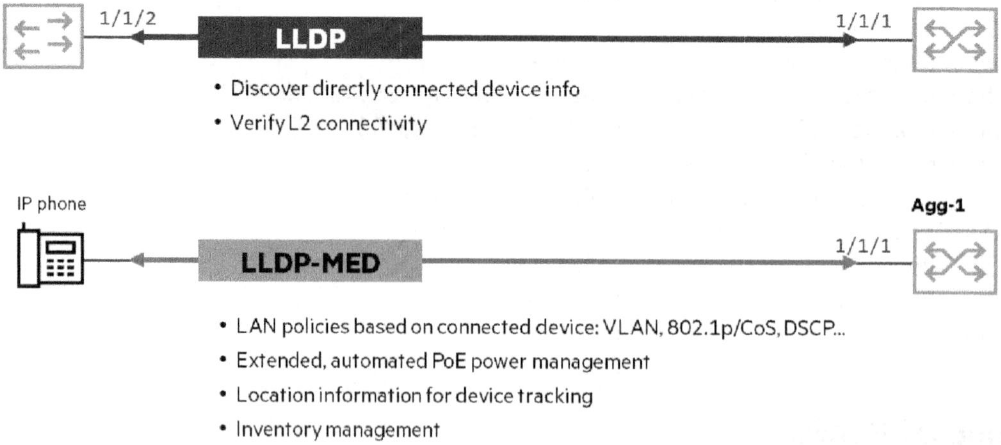

Figure 2-4: LLDP and LLDP-MED

Internet Control Message Protocol

Internet Control Message Protocol (ICMP) is a supporting protocol in the Internet protocol suite. Network devices use ICMP to send error messages and operational information, to indicate success or failure when communicating with another IP address. For example, an error is indicated when a requested service is not available or a host or router could not be reached.

Suppose that a new printer is added into the network and a laptop wants to use this device. The first troubleshooting step is to verify connectivity between these two devices, to ensure that the network is properly configured. ICMP helps you to address this problem.

What is ICMP?

- A support protocol in the TCP/IP suite.
- Devices exchange error messages and operational information.
- Helps validate network reachability.
- IP protocol number 1.

The following table shows the common messages that are used in ICMP.

Table 2-2: ICMP table message and description

Message	Description
Destination unreachable	Inform source of delivery issue
Time exceeded	TTL expired, packet discarded
Redirect	Inform others of a better L3 path
Echo request/reply	Ping: validate connectivity

ICMP messages do not include any TCP or UDP headers. ICMP messages are placed directly into an IP packet. Thus, ICMP is often considered to be an extension of the Network Layer.

Ping

The `ping` command is a utility used in networking to test the reachability of a host or device on an IP network and measure the round-trip time for data packets to travel between the source and destination. When the `ping` command is executed, the sending device sends a series of ICMP echo request packets to the destination host. If the destination host is reachable and responsive, it replies with ICMP echo reply packets, confirming the successful communication. The `ping` command provides valuable information about network connectivity and helps diagnose network connectivity problems, such as packet loss or high latency, by reporting the success or failure of packet transmission and measuring the time taken for a response.

In AOS-CX, the `ping` command must specify the IPv4 or hostname.

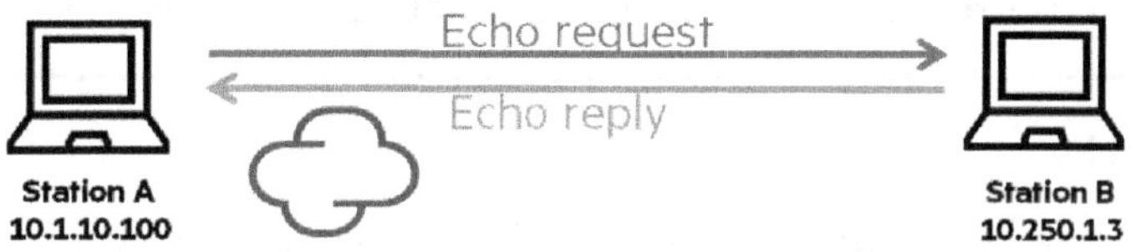

Figure 2-5: Ping echo request and rep

Syntax: `ping` *`<IPv4-ADDRESS-OR-HOSTNAME>`*

```
c:\Users\Administrator\ping 10.250.1.3
```

The `ping` command is supported in most client operating systems.

Traceroute

Traceroute is a network diagnostic tool used to trace the route that data packets take from the source to a specified destination across an IP network. When the traceroute command is executed, it sends a series of ICMP packets with varying TTL (Time-to-Live) values to the destination. Each router along the path decrements the TTL value, and when it reaches zero, the router sends an "ICMP Time Exceeded" message back to the source, allowing traceroute to identify the IP address and round-trip time to reach each router along the path, providing valuable insights into network latency and routing issues.

In AOS-CX, the `traceroute` command must specify an IPv4 address or hostname.

Syntax: `traceroute <IPv4-ADDRESS-OR-HOSTNAME>`

```
Access-1# traceroute 10.11.12.104
```

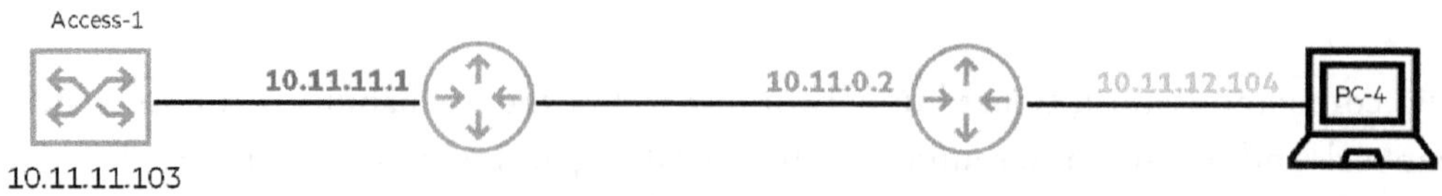

Figure 2-6: Traceroute example

In this figure, the switch tries to establish connectivity to the destination 10.11.12.104 via two intermediate routers (10.11.11.1 and 10.11.0.2).

Many firewalls block `ping` and `traceroute` commands. If you are attempting to use `ping` and `traceroute` commands to test connectivity through a firewall, the tests may fail. This does not necessarily mean that anything is wrong—it just means that the firewall is dropping your `ping` and `traceroute` commands as a security measure.

Power over Ethernet

Power over Ethernet (PoE) is a feature implemented on network devices such as Layer 2 and Multi-Layer switches, which provides power to endpoints using an Ethernet cable. This feature eliminates the need for external power sources, saving expenses on materials and installation time.

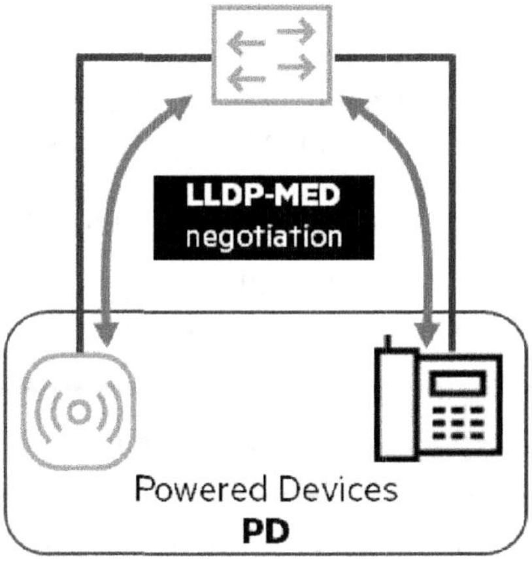

Figure 2-7: PoE negotiation

PoE uses the twisted pairs of a UTP cable to send power to PoE-enabled devices. The first PoE standard uses two twisted pairs to transmit data while the other two are used for power transmission. With the new PoE standards, power and data are sent over the four twisted pairs. If you are wondering how that is possible, the answer is that power and data use different frequencies, Electricity uses low frequencies (60 Hz or less), and data uses high frequencies (10-100MHz).

The standard considers two types of devices. Powered devices (PDs) are devices that receive PoE power, while Power Sourcing Equipment (PSE) provides the power. Typical PDs are access points, IP phones, cameras, and various IoT devices. On the other hand, network switches are consider PSEs.

There are four different standards that are available in the industry.

Table 2-3: PoE standards

Name	IEEE Standard	Class	Power to PD	Maximum PSE power per port	Energized pairs
PoE	802.3af	0 to 3	19.95 W	15.4 W	2-pairs
PoE+	802.3at	4	25.5 W	30 W	2-pairs
PoE++	802.3bt (Type 3)	5 and 6	51 W	60 W	4-pairs
PoE++	802.3bt (Type 4)	7 and 8	73.1 W	100 W	4-pairs

Learn more at https://www.arubanetworks.com/techdocs/AOS-CX/10.11/HTML/monitoring_6200/Content/Chp_PoE/pow-ove-eth-oveview.htm.

Checkpoint configuration

To create a checkpoint use the copy command as shown below:

```
Switch# copy {running-config | startup-config} checkpoint <CHECKPOINT-NAME>
```

To verify the checkpoint list, use the `show checkpoint` command.

To revert the switch from a checkpoint use the following command:

```
Switch# copy checkpoint <CHECKPOINT-NAME> {running-config | startup-config}
```

> You can export a checkpoint file to an external server.

Alternatively, to rollback to a checkpoint, you can use the `checkpoint rollback <CHECKPOINT-NAME>` command. This command only restores the configuration to the running-config file.

> The `checkpoint rollback` command does the same task as the `copy checkpoint running-config` command.

Checkpoint auto-mode

There are situations when applying a configuration could lock out your access to the switch. This leaves no other option but to locally connect to the device using the console port and remove the offending configuration lines.

The checkpoint auto mode feature addresses this problem in an elegant way. The solution includes a timer that you set. The system then creates a checkpoint with the current state, you can then make your desired configuration changes, which the system applies, after a confirmation. If the configuration lines did not affect your access, you will be able to confirm the change. If your changes locked you out of the device, confirmation does not occur. The switch thus automatically reverts to the latest snapshot after the time expires.

To use this mode, issue the following commands:

```
Switch# checkpoint auto <MINUTES>
Switch# checkpoint auto confirm
```

Switch provisioning options

Switch provisioning refers to the process of configuring and preparing network switches for deployment within a network infrastructure. This includes tasks such as setting up switch management interfaces, configuring VLANs, assigning IP addresses, enabling necessary protocols, and applying security policies.

There are several ways to provision AOS-CX switches, allowing network engineers to choose the best for their needs and scenarios.

- Manual provisioning
- Zero Touch Provisioning
- Fabric Composer
- HPE Aruba Networking Central / HPE Aruba Networking Central On-Premises

Manual provisioning

Manually provisioning a switch or a group of switches is a task that network engineers have been doing for decades. As a manual process, the network engineer needs physical access to the devices to connect to the console and/or management ports. Once connected, they need to perform device verification, show the power supply status and fan status, and check for any fault or unexpected output or error. Once the device is verified, they may start to update the switch software to the latest version or the one homologated by the customer. With the device tested and updated, they may begin configuring the switch.

The first step is to ensure that all the switch hardware is functional and ready, verifying the power supplies, fans, management modules (on chassis), and transceivers. During manual provisioning, you will need to update the switch software and configure the switch. Manual provisioning is a time-consuming task. To optimize the process, you may create the configuration scripts in advance and download the required software.

Initial setup - connectivity

For the initial configuration of an AOS-CX switch, three options can be used:

Console port

A console port is available on every AOS-CX switch. Keep in mind that they may vary in format.Not all Aruba AOS-CX switches have both RJ-45 console and USB-C console ports. The availability of console ports can vary depending on the specific model of the AOS-CX switch. Some AOS-CX switches may have both RJ-45 console and USB-C console ports, while others may only have one type of console port or may not have a console port at all. It's important to check the specifications of the specific AOS-CX switch model to determine the type of console ports available.

Once physically connected, you will need terminal emulation software to connect to the switch. The connection procedure is as follows:

- Connect the console port on the switch to the serial port on the computer using a console cable.
- Start the terminal emulation software on the computer and configure a new serial session with the following settings:
    - Speed: 115200 bps
    - Data bits: 8
    - Stop bits: 1
    - Parity: None
    - Flow control: None
- Start the terminal emulation session.
- Press **Enter** once. If the connection is successful, you are prompted to log in.

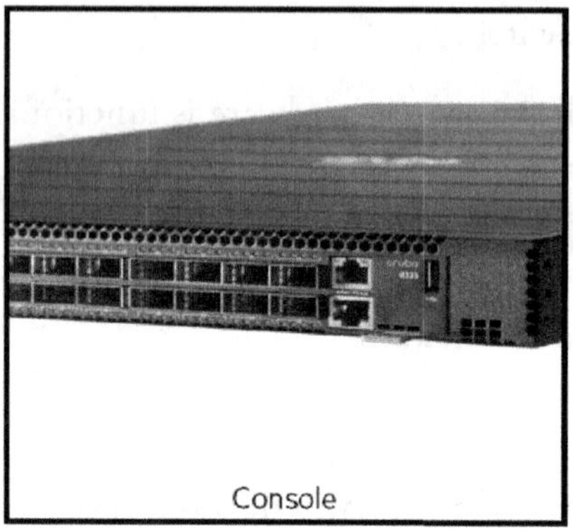

Figure 2-8: Console port

Management port

AOS-CX switches are equipped with Ethernet management interfaces that can be used for remote switch management. In data centers, this interface is often connected to the OOBM network. The management port can also be used for initial switch setup. By default, the management interface is set to obtain an IP address from a DHCP server automatically, and SSH support is enabled.

If there is no DHCP server on your network, you must configure a static address on the management interface. To connect to the switch using the management port, use an Ethernet cable to connect the management port to your network. Use another Ethernet cable to connect your computer to the same network. Then, start your SSH client software and configure a new session using the address assigned to the management interface. If the management interface is set to operate as a DHCP client, retrieve the IP address assigned to the management interface from your DHCP server. Start the session. If the connection is successful, you are prompted to log in.

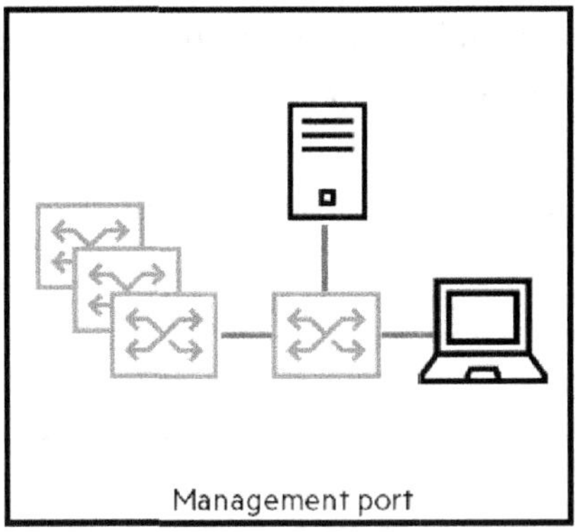

Figure 2-9: Topology of management port and how it connects to the switches

AOS-CX mobile app

The AOS-CX mobile app is an easy way to perform the initial switch configuration. Using the app, which is available on Android and Apple iOS, the network admin can use their smartphone to configure the management port, change the hostname, configure a switch stack, and much more. To use the mobile app, you need to connect a Bluetooth dongle to the switch USB-A port.

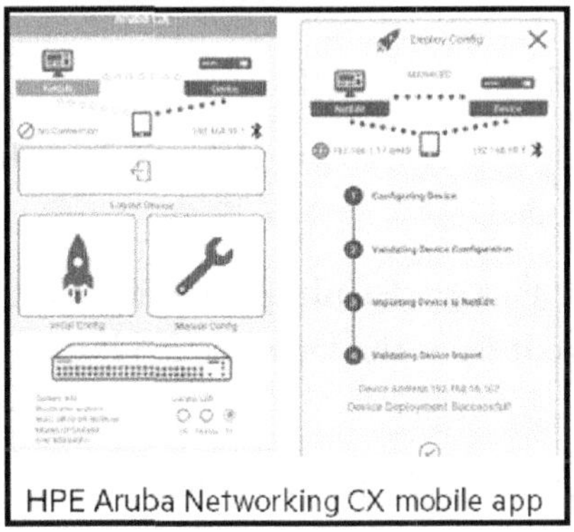

Figure 2-10: AOS-CX mobile app provisioning screens

Zero Touch Provisioning

Zero Touch Provisioning (ZTP) revolutionizes the deployment process of network devices by automating the configuration and setup tasks, eliminating the need for manual intervention. With ZTP, devices can be preconfigured with minimal initial settings and then automatically provisioned and configured once connected to the network. This approach streamlines deployment workflows, reduces human errors, and accelerates time-to-service activation, making it an indispensable tool in modern network management.

AOS-CX switches in their factory default configuration receive instructions from the DHCP server to download a specific software version and configuration file from a remote server. This process reduces the time and effort to configure a switch, since there is no need for the network admin to personally connect to each switch console to configure and update the software.

ZTP with DHCP workflow

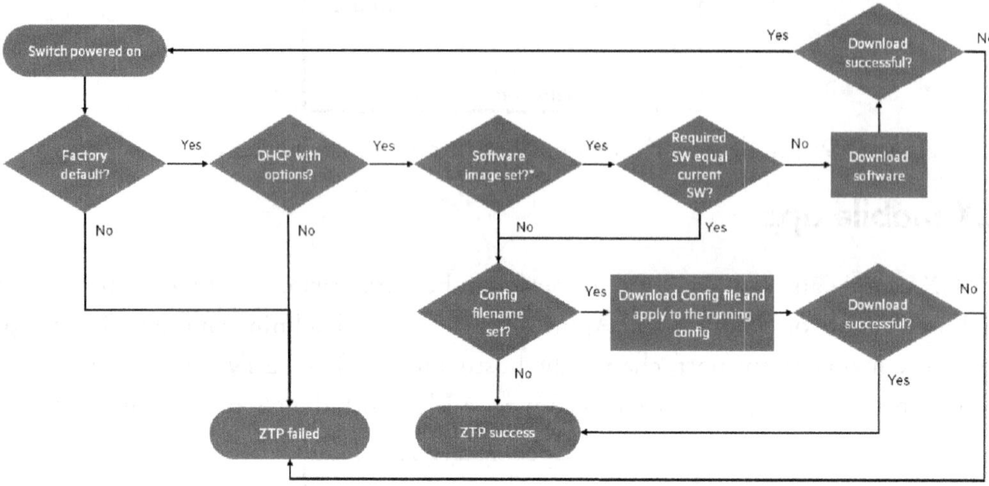

Figure 2-11: ZTP workflow process

The ZTP process looks like a simple process from the network admin standpoint, but under the hood, a series of processes are needed.

When a switch boots up with the factory default configuration, if the ZTP operation detects that the switch configuration is different from the factory default configuration, the ZTP operation ends.

The switch sends out a DHCP discovery from the management port. The switch waits to receive DHCP options indefinitely or until the running configuration is modified. If a DHCP IP address is received but no DHCP options are received, the switch waits an additional minute before ending the ZTP operation.

After the ZTP operation ends, there is no automatic retry. You can either attempt to boot the switch with the factory default configuration again, configure the switch at the installation site, or use the ZTP force-provision CLI to trigger the ZTP process, ignoring the present running configuration of the switch.

- Once force-provision is enabled, new DHCP requests are sent from the switch. Disabling force-provision does not stop the DHCP already in progress, but only changes the switch configuration status of force-provision.
- If ZTP fails while force-provision is enabled, there is no automatic retry. To retry, use the `ztp force-provision` command to disable and re-enable ZTP to clear the current ZTP state and send a new DHCP request. When `ztp force-provision` is already enabled on the switch, re-enabling it results in no operation.
- If the DHCP server is configured to provide both ZTP image and configuration options and there is a non-default startup configuration present on the switch, clearing the non-default startup configuration before triggering `ztp force-provision` is recommended. If an image is downloaded via ZTP, the switch reboots once the image download is complete and ZTP force-provision configuration is lost, causing ZTP to enter into a failed state. `ztp force-provision` will need to be enabled again to continue the process.

The DHCP server responds with an offer containing the following:

- The IPv4 address of the TFTP server
- One or both of the following:
    - The name of the software image file
    - The name of the configuration file
- HPE Aruba Networking Central Location (optional) including the shared token value for the on-premises server
- HTTP Proxy Location (optional)

If a software image file is offered, the ZTP operation downloads the image file from the TFTP server to the switch. If the current switch image and downloaded software image version do not match, then the switch boots with the downloaded image.

- If the image upgrade fails, the switch retains its original software image and the ZTP operation ends with a failed status.
- If the image upgrade succeeds, the ZTP operation is started again after the switch reboots. Because the downloaded image file matches the image file installed on the switch, the ZTP operation continues and checks if a configuration file is offered.

If a configuration file is offered, the ZTP operation downloads the configuration file and copies the file to the running-config, and then to the startup-config of the switch.

- If the startup configuration update fails, the switch retains its factory-default running configuration and the ZTP operation ends with a failed status.
- If the copy operation fails, the ZTP operation ends with a failed status.

If the copy operation succeeds, the ZTP operation ends successfully.

ZTP with DHCP

Provisioning switches with ZTP provides agility and reduces the time needed to provision new switches. But there are some preparation steps to be done ahead of time to enable new switches to receive software and configuration files. A TFTP server and the DHCP server must be configured. The DHCP server will provide the IP address and the TFTP server IP address through DHCP options and the name of the software and config file for download.

The DHCP server needs to be configured with DHCP options 43, 66, 67, and a series of sub-options, allowing the switches to learn the TFTP server IP address and file names for download from the DHCP server.

With the DHCP server configured, you need to download the desired software version from ASP and place it in the TFTP server "shared folder." Configuration files must be created for each switch and stored in the TFTP server shared folder.

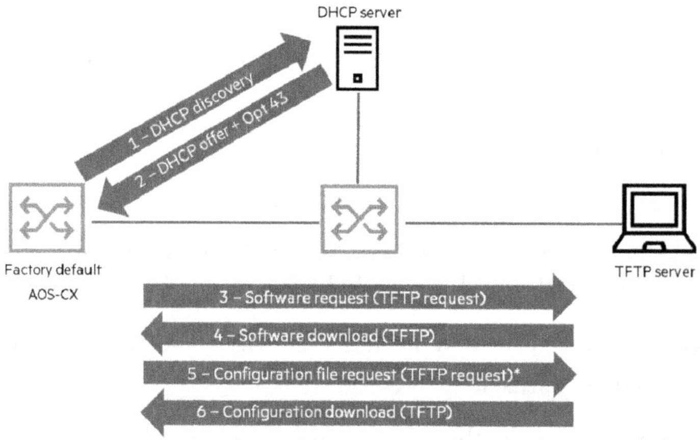

Figure 2-12: Topology flow of ZTP with DHCP

By default, an AOS-CX switch, with its factory default configurations, is configured as a DHCP client in the management interface. Once the switch powers up, it will send DHCP discovery packets. Once a DHCP server replies to the switch, it will verify if additional options such as 43, 66, and 67 were sent. If so, the switch will use this information to reach the TFTP server. First, if the DHCP server sends the software name, the switch will download the new firmware and reboot. If the DHCP server did not send the software name or the switch already has the same software version, it will request the configuration file and apply the configuration as both running and startup-config.

Supported DHCP options:

- 43 - Vendor Specific Information
- 43 Suboption 144 - Name of the configuration file
- 43 Suboption 145 - Name of the firmware image file
- 43 suboption 146 - FQDN or IPv4 address of the HPE Aruba Networking Central On-Premises server
- 43 suboption 148 - FQDN or IPv4 address of the HTTP Proxy
- 60 - Vendor Class Identifier (VCI)
- 66 - IPv4 address of the TFTP server (Specifying a host name instead of an IP address is not supported.)
- 67 - Name of the configuration file (Option 43 suboption 144 takes precedence over this option.)

* The configuration file is a text file or JSON file that becomes the startup and running configuration on the switch after the ZTP operation is complete. The configuration can be in CLI or in JSON format.

Fabric Composer

Fabric Composer is designed to work with AOS-CX switches to optimize switch and fabric provisioning seamlessly. Using a series of interactive and automated workflows, Fabric Composer allows network administrators to provision data center switches easily.

Fabric Composer orchestrates a discrete set of switches as an entity in a "fabric" topology, simplifying day-to-day operations and troubleshooting.

Fabric Composer supports deep IT ecosystem integrations that enable administrators to manage, provision, and visualize their entire end-to-end network. An intuitive HTML5-based user interface provides a self-documenting real-time map of the environment—from a VM to host networking, to a specific switch port, and through Networks View visualizing Virtual Switch Extensions and spine-leaf connectivity. A robust set of APIs and tight integration into many third-party data center automation stacks provide a seamless configuration and operational experience.

HPE Aruba Networking Central

HPE Aruba Networking Central is a powerful cloud-managed networking solution that offers unmatched simplicity for today's IT operations. As the management and orchestration console for the HPE Aruba Networking Edge Services Platform, HPE Aruba Networking Central provides a single point of control to oversee every aspect of wired and wireless LANs, WANs, and VPNs across campus, branch, remote, and data center locations.

AI-powered analytics, end-to-end orchestration and automation, and advanced security features are built natively into the solution. Live upgrades, robust reporting, and live chat support are also included, bringing more efficiency to day-to-day maintenance activities. The option to use ZTP to onboard new switches to the platform brings agility when adding new switches and devices.

HPE Aruba Networking Central can be run as a cloud-native, microservices architecture, delivering enterprise requirements for scale and resiliency. The on-premises option is available for customers and environments that cannot have internet access, such as highly secure verticals (defense forces, banks).

Switch monitoring

The `show` CLI explores invaluable commands used to retrieve detailed information about the configuration and status of networking devices. These commands provide administrators with real-time insights into device operation, facilitating troubleshooting, monitoring, and maintenance tasks with precision and efficiency.

Port numbering

Fixed versus modular

AOS-CX reference interfaces following the member/slot/port notation:

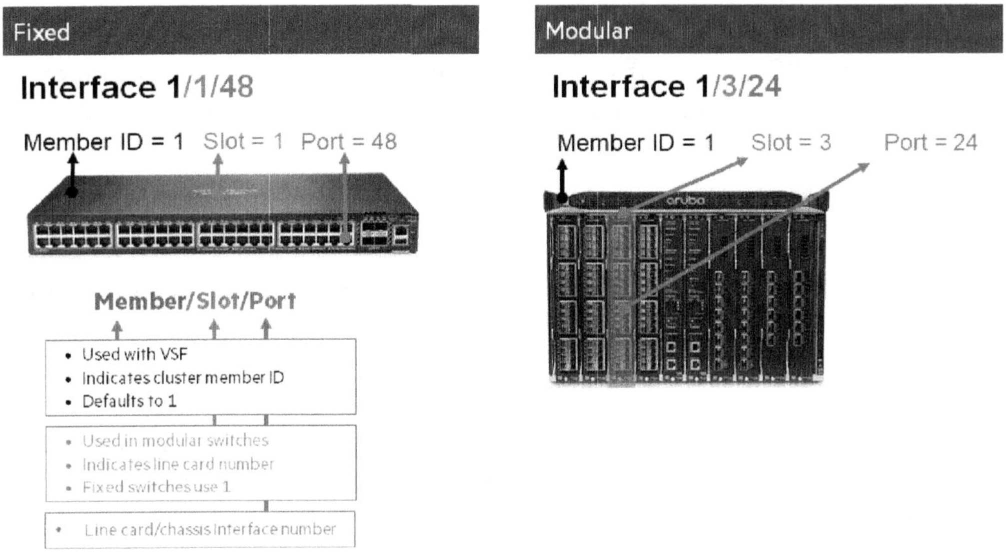

Figure 2-13: Example with a fixed switch and a modular switch

Member: This member number indicates the member ID in the cluster. By default, an AOS-CX switch will be member 1, indicating that these features are not running.

Slot: In modular switches like the CX 8400 Series and CX 6400 Series, this number represents the slot being used by a particular line card. For fixed switches (CX 8320 Series, CX 8325 Series, and CX 6300 Series), this number will be always 1.

Port: This number refers to the individual interface in the line card (modular switches) or in the chassis (fixed switches).

Show commands

There are basic commands that will help you to verify your networking equipment, along with its general capabilities. This is some of the most valuable information you can learn during an entry-level networking course. Good network engineers and administrators must be adept at discovering the actual connectivity of the network. This will help you to verify the accuracy of existing documentation, or to create new network documentation from scratch.

These `show` commands are also valuable for baselining and troubleshooting. Using these commands, you can view many switch performance and health parameters, including CPU and memory utilization. Suppose that your network is healthy, everyone is happy, and all your switches are generally at about 5% CPU utilization. You have been paying attention to this for months now—you know your network.

During an outage or slowdown, you find a switch that is running at 85% utilization. You might focus your diagnostic efforts on that switch and its directly attached devices. Because you paid attention while the network was healthy, you are more efficient when the network is not healthy.

If you had not paid attention, you would have no idea whether 85% utilization is normal or abnormal. Of course, HPE has very effective management platforms that ease and automate these baseline processes, but it can still be valuable to use the CLI to check network device health and status.

Mastering the commands that you are about to learn, creating good documentation, and having good baseline information is kind of like getting the answers to a test before you take the test. Only this is the real world, and the "test" relates to how effective you are when the network is down.

show version

The `show version` command displays version information about the network operating system software, service operating system software and BIOS. In the CLI, you see that this system is running the AOS-CX, version GL.10.04.0003. Near the bottom you see the Service OS and BIOS versions. This is useful for your general knowledge, and to help to determine if upgrades are needed.

```
Switch# show version
-----------------------------------------------------------------
AOS-CX
(c)Copyright 2017-2019 Hewlett Packard Enterprise Development LP
-----------------------------------------------------------------
Version : GL.10.13.1000
<output omitted for clarity>
Service OS Version          : GL.01.14.0001
BIOS Version                : GL-01-0013
```

show system

The `show system` command displays general status information about the system. During remote access, the device platform and version information may not be obvious. In the CLI displayed, you are attached to the device named "Switch."

A CX 8325-48YC has 48 25Gbps ports, and eight 100Gbps ports. You see the serial number, base MAC address, and uptime, as well as CPU and memory utilization. This can give you some idea of the general health of the system. By baselining these parameters, you can begin to establish what "normal" is for your network.

```
Switch# show system
Hostname                    : Switch
System Description          : GL.10.13.1000
System Contact              :
System Location             :
Vendor                      : Aruba
Product Name                : JL635A Aruba 8325-48Y8C 48p 25G 8p 100G swch
```

```
Chassis Serial Nbr         : TW98KM003P
Base MAC Address           : 9020c2-bced00
AOS-CX Version             : GL.10.13.1000
Time Zone                  : UTC
Up Time                    : 1 week, 1 day, 11 hours, 20 minutes
CPU Util (%)               : 4
Memory Usage (%)           :19
```

top cpu

The top cpu command shows detailed CPU utilization information. You learned how to see high-level CPU utilization with the show system command. The top cpu command provides more detailed information.

```
Switch# top cpu
top - 20:40:46 up 17 days, 23:45, 1 user, load average: 0.32, 0.26, 0.24
Tasks: 238 total, 1 running, 194 sleeping, 0 stopped, 0 zonbie
%Cpu(s): 2.5 us, 1.6 sy, 0.0 ni, 95.9 id, 0.0 wa, 0.0 hi, 0.0 si, 0.0 st
KiB Mem : 7868164 total, 3013360 free, 13999328 used, 3455476 buff/cache
KiB Swap: 2097148 total, 2095868 free, 1280 used. 6079420 avail Mem
PID     USER     PR    NI    VIRT     RES     SHR       S    %CPU    $MEM    TIME+
   COMMAND
26308   admin    20    0     3344     2064    1612      R    11.8    0.0     0:00.03
   top -n -n 1 -s -c -o %CPU
629     root     20    0     33928    17084   6124      S    5.9     0.2
777:57.14   /usr/sbin/ovsdb-server
3047    root     20    0     791184   84416   32232     S    5.9     1.1
189:43.13   /usr/bin/prometheus
1       root     20    0     7600     6312    2960      S    0.0     1.1     0:28.14
   /usr/bin/prometheus
```

top memory

The top memory command shows memory utilization information.

```
Switch# top memory
top - 20:41:22 up 17 days, 23:46, 1 user, load average: 0.31, 0.26, 0.24
Tasks: 238 total, 2 running, 193 sleeping, 0 stopped, 0 zonbie
%Cpu(s): 2.5 us, 1.6 sy, 0.0 ni, 95.9 id, 0.0 wa, 0.0 hi, 0.0 si, 0.0 st
KiB Mem : 7868164 total, 3013768 free, 1398760 used, 3455636 buff/cache
KiB Swap: 2097148 total, 2095868 free, 1280 used. 6080024 avail Mem
PID     USER     PR    NI    VIRT      RES     SHR       S    %CPU    $MEM    TIME+
   COMMAND
2366    admin    20    0     2013036   2064    45568     S    0.0     1.6
739:59.05   /usr/bin/switchd_agent -i 1 -s
3047    root     20    0     791184    17084   32232     S    0.0     1.1
189:43.37   /usr/bin/prometheus
```

```
701       root     20    0    1212616  84416  19316    S    0.0    1.0
18:02.19    /usr/bin/HPE-restd
```

show running-config

The `show running-config` command displays the current non-default configuration running on the switch. No user information is displayed. This provides you with a detailed snapshot of the device's operation parameters.

```
Switch# show running-config
Current configuration:
!
!Version AOS-CX GL.10.13.1000
!export-password:       default
Hostname Switch
!
Router ospf 1
area 0.0.0.0
Vlan 1,1101,1103,1105,1107
Interface mgmt
no shutdown
ip static 10.250.1.3/24
default-gateway 10.250.1.254
Interface 1/1/29
no shutdown
description To-8325-Core-Table01-Port47
no routing
vlan trunk native 1
vlan trunk allowed 1101,1103,1105
<lines omitted due to space limitation>
```

show events

The `show events` command allows you to view a chronological list of system events and activities logged by the switch. This command provides insight into the operational history, aiding in troubleshooting, performance analysis, and security incident investigations.

```
Switch# show events -r
-------------------------------------------------
Event logs from current boot
-------------------------------------------------
2019-12-06T10:24:18.679695+00:00 Spine-8325-C hpe-sysmond[3212]: Event|6303|LOG_
INFO|AMM|1/1|Current system memory usage is 27%
2019-12-06T09:09:19.067316+00:00 Spine-8325-C lldpd[2165]: Event|106|LOG_
INFO|AMM|1/1|LLDP neighbor 90:20:c2:bc:7c:00 deleted  on 1/1/35
```

The `show event` command has other parameters that you can use, such as `-e`, `-s`, `-a`, `-n`, `-c`, and `-d`. You can learn more about how to use these parameters in the *Command-Line Interface Guide* for your switch and software version.

show interface brief

The `show interface brief` command helps the administrator to see the available interfaces and the current status. This command also briefly shows Layer 2 and Layer 3 configurations applied to the interface. This is a frequently used command for many network engineers.

At a glance, you can tell the status of every interface on the device, with a specific focus on the Enabled and Status columns.

Enabled "yes" means that the interface is not disabled in the switch configuration—no administrator has disabled the interface.

"Up" in the status column means that something is attached to that interface, and there is at least a good Layer 1/Layer 2 connection between them.

```
Switch# show interface brief
----------------------------------------------------------------------
Port      Native  Mode    Type         Enabled Status  Reason    Speed
          VLAN                                                   (Mb/s)
----------------------------------------------------------------------
1/1/1     1       routed  SFP28DAC3    yes     up                25000
1/1/2     1       trunk   SFP28DAC3    yes     up                25000
1/1/3     1       trunk   SFP28DAC3    yes     up                25000

<lines omitted due to space limitation>
```

show interface transceiver detail

The `show interfaces transceiver detail` command lists the transceivers installed in the switch.

```
Switch# show interface transceiver detail
Transceiver in 1/1/1
Interface Name       : 1/1/1
Type                 : SFP28DAC3
Connector Type       : No separable connector
Transfer Distance    : 0.00km (SMF), 0m (OM1), 0m (OM2), 0m (OM3)
Diagnostic Support   : None
Product Number       : JL488A
Serial Number        : CN97KBG0BJ
Part Number          : 8121-1743
```

Learning check

Check your understanding of this material by answering the following questions.

Question 1

What kind of cables might you use to connect to the AOS-CX switch console port? (Select two.)

- ☐ Ethernet cable
- ☐ Fiber optic cable
- ☐ USB cable
- ☐ Serial cable

Question 2

Which command could you use to validate the status of the interfaces on the switch?

- ○ `show running-config`
- ○ `show system`
- ○ `show interface brief`
- ○ `show events`
- ○ `show interface transceiver detail`

Question 3

Which of the options below accurately describe network discovery commands or techniques?

- ○ LLDP is a Layer 2 discovery protocol.
- ○ You typically use the `ping` command to help document your network.
- ○ If the `ping` or `traceroute` command fails, you know that you have connectivity issue.
- ○ The `ping` command leverages ICMP echo requests and echo replies.

3 VLANs

Welcome to the Virtual Local Area Networks (VLANs) chapter, where you will understand the benefits of VLANs and the steps involved in configuring some of the VLAN features. This chapter will introduce you to 802.1Q frame tagging. You will learn about collision and broadcast domains, MAC addresses, ARP forwarding tables, and the frame delivery process.

OBJECTIVES

After completing this chapter, you should be able to:

- ✓ Identify collision and broadcast domains.
- ✓ Demonstrate the configuration of VLANs.
- ✓ Analyze 802.1Q tagging.
- ✓ Implement trunk ports.
- ✓ Differentiate MAC addresses and ARP forwarding tables.
- ✓ Evaluate the frame delivery process.

Domains - Collision and broadcast

Domains refer to logical groupings of devices within a network based on shared characteristics or requirements. VLANs create separate broadcast domains, preventing unnecessary broadcast traffic from reaching devices outside the VLAN. By isolating traffic within specific domains, VLANs enhance network security, efficiency, and management capabilities.

Let us start this chapter with a simple concept: the domain. In computer networking, a domain refers to a system of interconnected network devices and network objects. A collision occurs when two devices on the same domain send data packets at the same time.

- **Collision domain:** A collision domain is a part of a network where data packets can collide with one another when transmitted simultaneously, leading to data corruption and performance degradation.

- **Broadcast domain:** A broadcast domain is a part of the network segment where devices at all nodes can reach each other through a broadcast.

Segregating these domains using VLANs helps reduce collisions, control broadcast traffic, and enhance network performance and security.

Collision domains are a natural occurrence in shared network environments where multiple devices communicate over a shared medium, such as in an Ethernet network. They emerge from the physical limitations of how data transmission works in these types of networks, where collisions may happen when multiple devices try to send data simultaneously.

To address collision issues and improve network performance, network administrators can intentionally design and segment the network into separate collision domains using techniques like VLANs. By doing so, they can control the flow of traffic and reduce the likelihood of collisions, while also managing broadcast domains to limit unnecessary broadcast traffic and improve network efficiency.

Collision domain

Let us understand the collision domain concept with the help of a simple scenario. Suppose Alice and Bob are having a conversation. Alice has something to say, but her ears detect that Bob is also speaking, so she politely waits her turn. If Alice and Bob happen to speak at the same time, the sounds of their voices collide, making it difficult to understand. Alice and Bob realize what is happening and stop. Bob says, "I'm sorry, please go ahead and speak." Alice says, "No, please, you go ahead." They both back off and then try again.

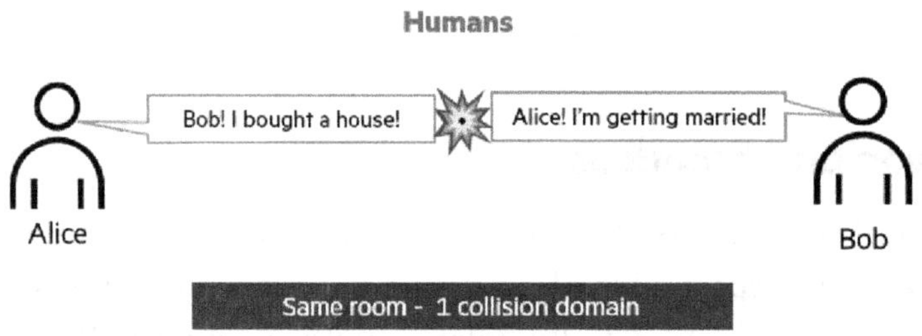

Figure 3-1: Two people trying to talk at the same time, which is an example of collision domain

Collision domain: Wired

In a wired network scenario such as Ethernet, only one Ethernet host on a particular segment may transmit at a time. This is all controlled by an algorithm running on all Ethernet Network Interface Cards (NICs), which is known as Carrier Sense Multiple Access / Collision Detection (CSMA/CD). This works very much like humans, as described in the previous example.

Multiple humans can be in the same small room, and all of them can access the airwaves and speak any time they want; this is a multi-access system, like Ethernet. Before speaking, however, a polite human first listens to sense if others are currently speaking, which is just like Ethernet's Carrier Sense mechanism. Ethernet cards sense the state of a "carrier signal" that indicates a currently transmitting NIC.

If two devices transmit at the same time, the electrical signals collide and become corrupted. This is called a collision. This condition is detected, and each station backs off and tries again.

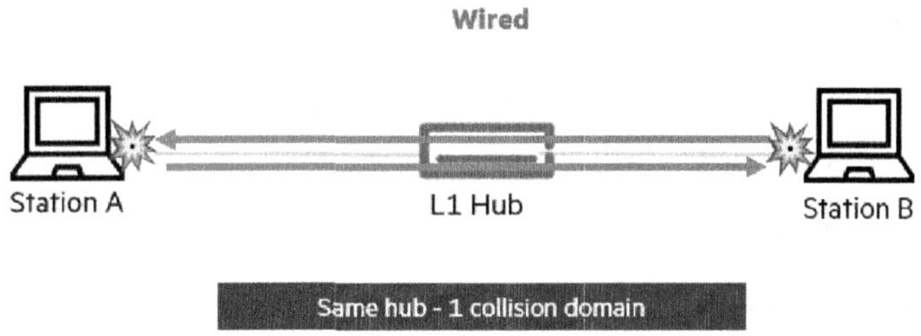

Figure 3-2: Two computers connected to the same hub

Ethernet collisions occur when a hub is deployed in network. A hub is a Layer 1 device that receives digital ones and zeros, and repeats them, near instantaneously, out all other ports. Therefore, a hub is also known as a multi-port repeater. All 16 devices connected to a 16-port hub are in the same collision domain; that is, if one device transmits, 15 other devices must wait. If two hosts transmit at the same time, a collision occurs and everyone backs off, waits a few milliseconds, and tries again.

However, hubs are outdated devices that are no longer used. These unintelligent, Layer 1 repeaters have been replaced by intelligent Layer 2 switches. For a properly deployed and configured switch under normal circumstances, collisions do not occur on Layer 2 switches.

Collision domain: Wireless

Similar to a wired network, Wi-Fi uses an algorithm called Carrier Sense Multiple Access / Collision Avoidance (CSMA/CA). This works very much like Ethernet's CSMA/CD, only it adds other mechanisms to preemptively avoid collisions. Everyone attached to the same wireless access point (AP) channel is in the same collision domain. Therefore, only one of them may transmit at a time.

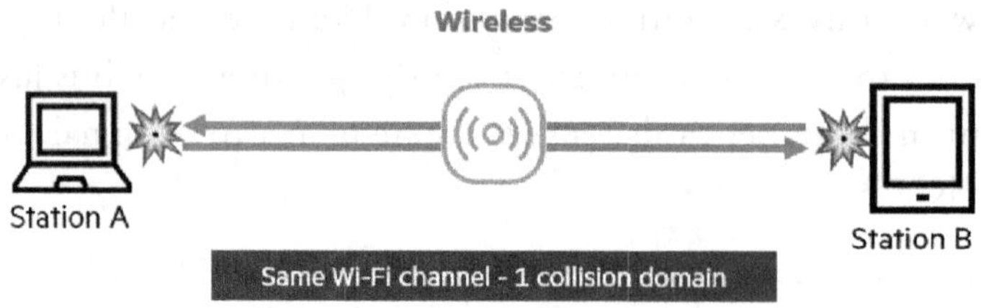

Figure 3-3: Two computers connected to the same wireless access point

Collision domains and performance

Your understanding of collision domains enables you to optimize network performance, which is explained in the following scenarios.

Scenario 1: Group conversation

Consider a group of people in a room. If everyone is talking about the same topic, the conversations will be easy and efficient. However, if there are two topics that are being discussed at the same time, then people will find it difficult to understand. People must also wait a long time for their turn. To optimize this scenario, you can split people into two rooms based on the topic they are discussing. This is similar to creating two collision domains.

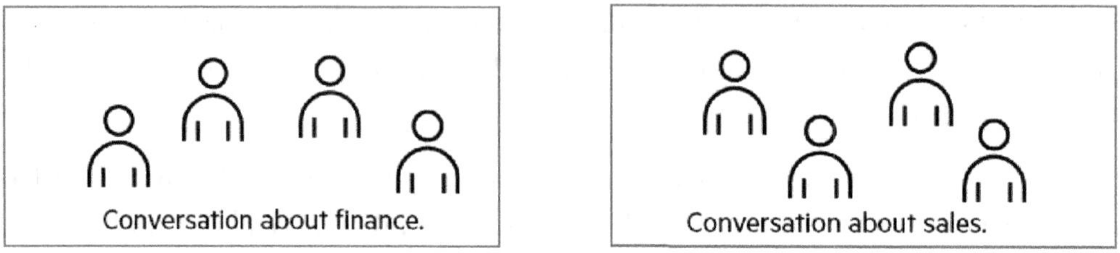

Figure 3-4: Group conversations

Scenario 2: Wired network

Suppose that you have an Ethernet hub with eight connected devices. While one host transmits, seven hosts must wait their turn. If you had a 48-port hub, then 47 hosts must wait—imagine the performance impact. To optimize this scenario, you would want to split this network up into multiple collision domains. Thankfully, hubs are relatively outdated devices that are no longer used. These unintelligent, Layer 1 repeaters have been replaced by more intelligent Layer 2 switches. For a properly deployed and configured Layer 2 switch under normal circumstances, collisions do not occur.

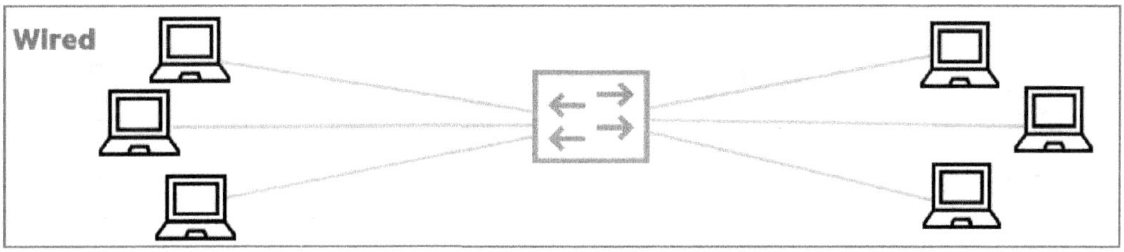

Figure 3-5: A wired network

Scenario 3: Wireless network

Consider an AP in a room, with the transmit power set to maximum to increases its coverage area. There are 50 people present in the room who are connected to the single AP. This may save money as only one AP is providing connection to all 50 hosts. However, while one host transmits, 49 people must wait; people will start complaining about the network performance. To optimize your network, you must purchase one more AP. Lower the transmit power of both APs and configure a unique channel to both APs. Now, 25 people will be connected to each AP, which acts as different collision domain. Two people can now transmit at the same time, improving the network performance.

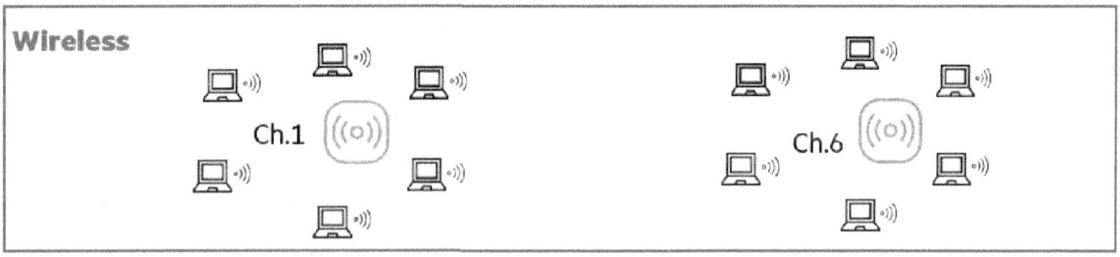

Figure 3-6: A wireless network

Broadcast domains

A broadcast domain is a group of devices that are on the same network, capable of receiving and responding to a broadcast frame from any device. A broadcast frame uses a special Layer 2 MAC address: FF:FF:FF:FF:FF:FF, and a broadcast packet uses a special Layer 3 IPv4 address: 255.255.255.255. Layer 2 switches always forward any received broadcast frame out all ports, except for the port on which the broadcast was received (the ingress port).

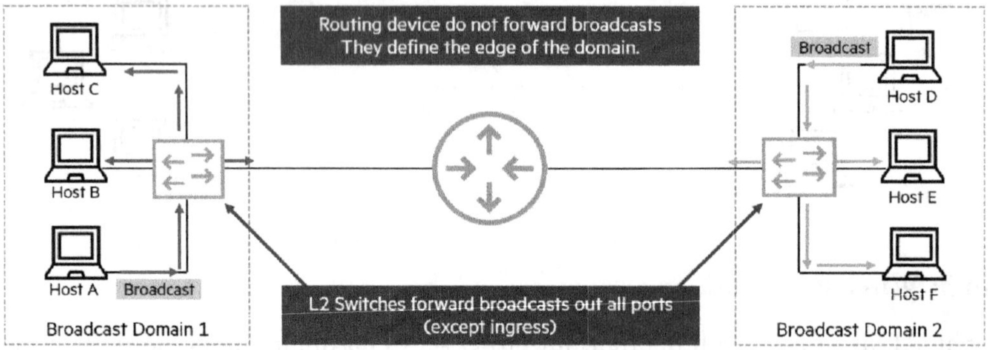

Figure 3-7: Broadcast domains

In this figure, host A transmits a broadcast frame. Perhaps it is saying, "Hey everyone! Who has IP address 10.1.1.1?" Because the switch forwards this frame out all other ports, hosts B, C, and the router receive the message. Nobody in broadcast domain 2 receives the message as they are on a different network. Similarly, when host D sends a broadcast, all hosts in domain 2 receive it, and nobody in domain 1 receives it.

Let us now connect these two domains with the help of a router. If host A sends a broadcast frame, it does not reach domain 2 and vice versa. Thus, the router defines the edge of the broadcast domain. You may be wondering, why not remove the router altogether and let the two domains be connected? This would save the cost of purchasing a router; however, it will cause performance issues. Broadcast traffic will flood the network, leaving no resources for valid traffic, resulting in Denial of Service (DoS). So, smaller broadcast domains mean better performance.

LANs and VLANs

LAN

A Local Area Network (LAN) is a group of devices in the same broadcast domain. **Figure 3-8** shows two LANs, labeled LAN-10 and LAN-20. The LAN-10 switch has two connected devices. Host A connects to port 1 and host B connects to port 2. The other switch has similar connections.

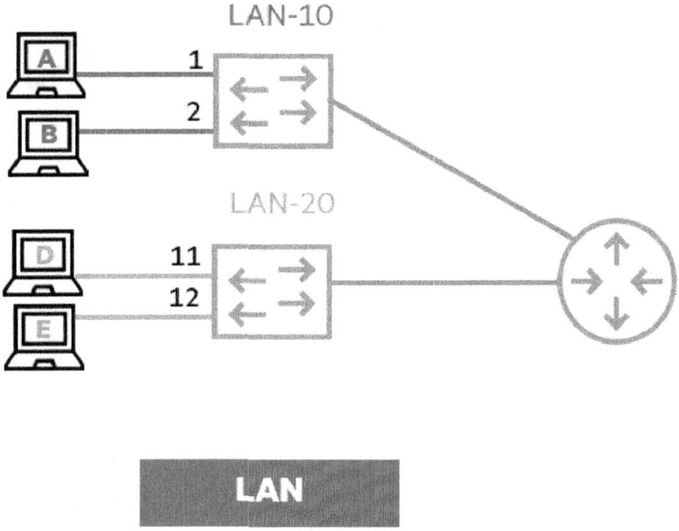

Figure 3-8: Example LANs

If a switch receives a broadcast on any port, it floods it out all other ports, except the ingress port. Therefore, if host A sends a broadcast, the LAN-10 switch forwards it out all other ports, and so host B receives the frame. Of course, hosts D and E do not receive this frame as there is no connection between LAN-10 and LAN-20. They are physically separate LANs, connected to physically separate switches. Now, if you want to connect these two LANs into an internetwork, you can add a router as shown. The router will route unicast and multicast traffic between the LANs.

Virtual LAN

Consider a Virtual LAN (VLAN), which is just like a physical LAN, a group of devices in the same broadcast domain. Suppose that you have a single, physical switch named SW1.

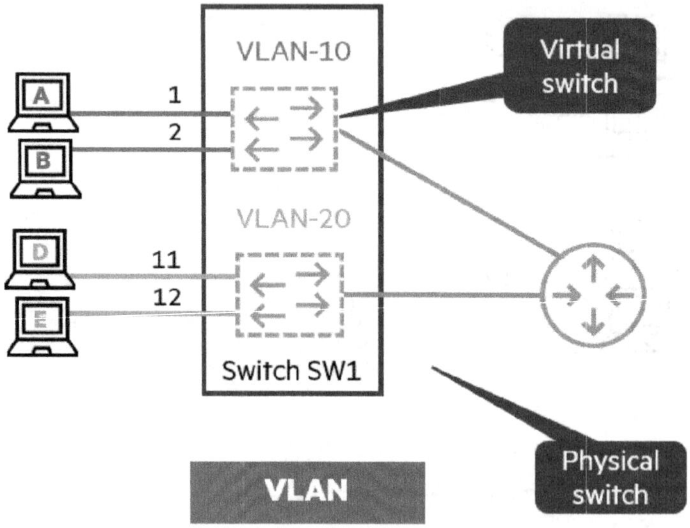

Figure 3-9: Virtual LANs example

Hosts A, B, D, and E connect to ports 1, 2, 11, and 12 on this switch. By default, all these devices are on VLAN-1, thus on the same broadcast domain. If host A sends a broadcast, all other hosts receive the frame. Now suppose that you created a Virtual LAN named "VLAN-10." It is as if you have created a small virtual switch inside the physical switch. This virtual switch exists, but it is not initially connected to any of the physical switch ports. So you define or "map" physical ports 1 and 2 as being members of VLAN-10.

Similarly, create VLAN-20, and assign ports 11 and 12 as members of VLAN-20. You have now created two separate broadcast domains on a single physical switch. When host A sends a broadcast, the switch knows to only forward that frame out all ports that are in the same VLAN, except the ingress port. Only host B receives the broadcast. If host D broadcasts, only host E receives it.

You have effectively re-created the initial scenario (LAN) at the beginning of this topic (before the router was added), using a single physical switch. There is no connectivity between the separate VLANs. No unicast, multicast, or broadcast traffic shall pass between the VLANs.

VLAN creation

Let us now learn how to create, disable, remove, and rename a VLAN in AOS-CX. In AOS-CX, VLAN 1 is created by default and cannot be removed. By default, all ports are initially members of this VLAN, a common default for many switches. AOS-CX refers to VLANs by their VLAN-ID, a number between 1 and 4094.

Create a VLAN

```
SW1(config)# vlan 10
SW1(config-vlan-10)#
```

Create multiple VLANS

```
SW1(config)# vlan 2-5,10
```

Remove VLAN 10

```
SW1(config)# no vlan 10
```

Disable VLAN 10

```
SW1(config)# vlan 10
SW1(config-vlan-10)# shutdown
```

Rename VLAN 10

```
SW1(config)# vlan 10
SW1(config-vlan-10)# name Sales
```

Remember, when you use the `vlan 10` command to create a VLAN, it is as if you have created a virtual switch inside of the actual, physical switch. Although it exists, no ports have yet been defined as members of this VLAN. No devices are attached to this virtual switch yet, as shown below. The next step after creating a VLAN is to map or associate physical interfaces as members of the VLAN.

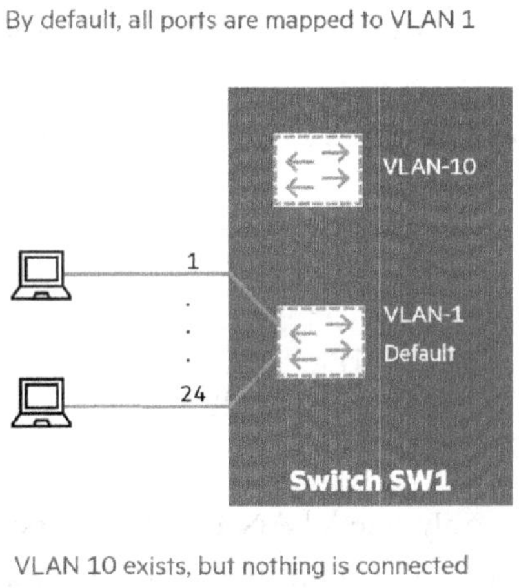

Figure 3-10: Ports mapped to a VLAN

Mapping ports to VLAN

You learned how to create a VLAN in the previous section. Now, let us understand the steps involved in mapping the ports of your switch to the VLAN. Let us define VLAN 10 and VLAN 20. We will assign the names Sales and Services to VLAN 10 and VLAN 20 respectively.

```
SW1(config)# vlan 10
SW1(config-vlan-10)# name Sales
SW1(config-vlan-10)# exit
SW1(config)# vlan 20
SW1(config-vlan-20)# name Service
SW1(config-vlan-20)# exit
```

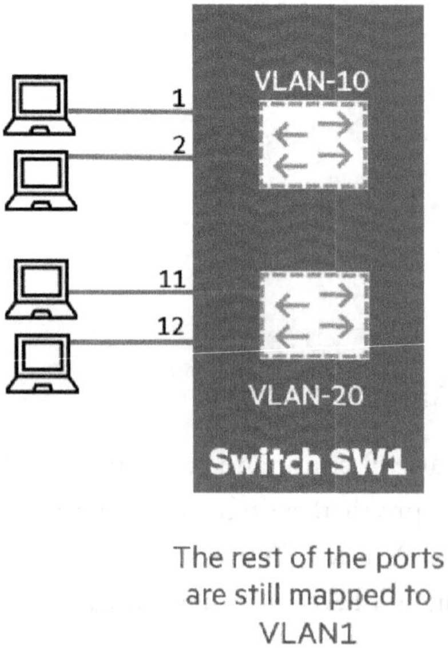

Figure 3-11: VLAN-10 and VLAN-20 mapped to different ports of a switch

Now, we will map (assign) ports to the VLANs. From the global configuration context, you can choose to configure a range of interfaces at the same time to ports 1/1/1 through 1/1/2 before mapping them to VLAN 10, shown as follows. Similarly, we will map ports 1/1/11 through 1/1/112 to VLAN 20.

```
SW1(config)# interface 1/1/1-1/1/2
SW1(config-if-<1/1/1-1/1/2)# vlan access 10
SW1(config-if-<1/1/1-1/1/2)# exit
SW1(config)# interface 1/1/11-1/1/12
SW1(config-if-<1/1/4-1/1/5)# vlan access 20
SW1(config-if-<1/1/4-1/1/5)# exit
```

It is important to understand that only one VLAN ID can be assigned to an interface. For example, interface 1/1/1 cannot be a member of both VLAN 10 and VLAN 20 at the same time. Let us use the `show vlan` command to verify the VLAN creation and mapping.

```
SW1# show vlan
---------------------------------------------------------------------
VLAN    Name              Status   Reason   Type     Interfaces
---------------------------------------------------------------------
1       DEFAULT_VLAN_1    up       ok       default  <output omitted>
10      Sales             up       ok       static   1/1/1-1/1/2
20      Service           up       ok       static   1/1/11-1/1/12
```

802.1Q

802.1Q is a standard for VLAN (Virtual Local Area Network) tagging in Ethernet networks. It is an extension of the IEEE 802.3 Ethernet standard and allows for the identification and segregation of different VLANs within a network. By adding a VLAN tag to Ethernet frames, 802.1Q enables the transmission of multiple VLANs over a single physical network infrastructure, facilitating network segmentation, improved security, and efficient traffic management. Before we start discussing the 802.1Q standard and the process, let us go through some basic definitions that will help you better understand these concepts.

Trunk

A trunk is a link between two switches that can connect multiple VLANS. Packets are encapsulated with VLAN identifiers, which are used to direct them to their respective VLANS.

Tagging

VLAN tagging is the method of encapsulating a packet travelling through a trunk link with a VLAN identifier.

Extending VLAN using 802.1Q tagging

Let us consider the same scenario as in the previous section. A standard Ethernet frame is shown in the following figure.

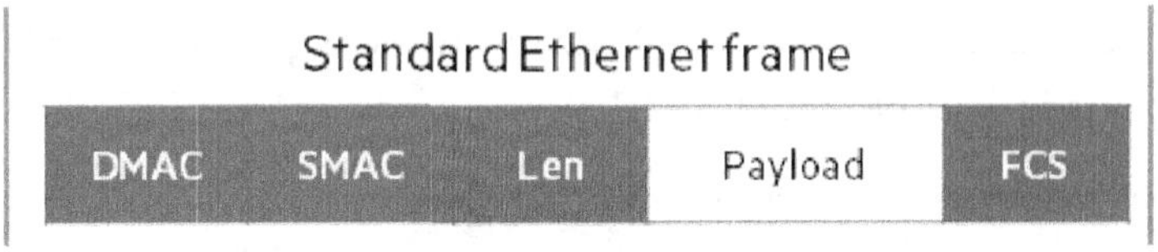

Figure 3-12: A standard Ethernet frame

The two switches are connected through a trunk link to carry VLAN traffic. If this Ethernet frame is a broadcast from host A, flooded out to all ports in VLAN 10, including port 24, there is no field in this frame to inform receiving switches about the intended destination VLAN. If SW2 could talk, it might ask, "What do I do with this? Is this for VLAN 10? VLAN 20? VLAN 1?"

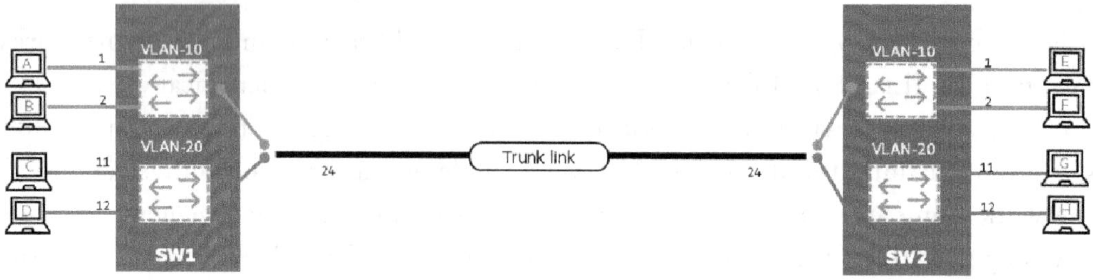

Figure 3-13: 802.1Q tagging

To address this issue, we need a tagging mechanism, which is defined in the IEEE 802.1Q standard. This standard adds an extra field between the length and payload fields: the 802.1Q Tag field. The most important part of this field is the VLAN tag or VLAN-ID, or simply VID. Before SW1 floods this broadcast frame across this specially defined "trunk port," it adds this tag to the frame, as follows.

Figure 3-14: VID 10 added to the frame

SW2 receives the frame, looks at the tag, and knows that this frame is for VLAN 10. It strips off the special tag and forwards a standard Ethernet frame out on all ports in VLAN 10 to hosts E and F. Next, host G sends a broadcast. SW2 floods a standard Ethernet frame out on its local ports. Before flooding the frame out port 24 , it adds a tag, "This frame is for VLAN 20." SW1 receives this frame, sees that it is for VLAN 20, strips off the tag, and forwards a standard Ethernet frame out all ports in VLAN 20 (11 and 12).

Figure 3-15: VID 20 added to the frame

Now, you can extend the VLAN across multiple switches, using a single physical port.

The VLAN-ID field is 12 bits long, which allows a switch to carry traffic for up to 4094 VLANs.

The 802.1Q standard allows administrators to select a single VLAN on which no tag or VID is included in the Layer 2 header. This means that a standard, untagged Ethernet frame is sent out of the port. This VLAN is known as the native VLAN or untagged VLAN. By default, in AOS-CX, VLAN 1 is the native VLAN. You can modify this by using the `vlan trunk native <VLAN-ID>` command.

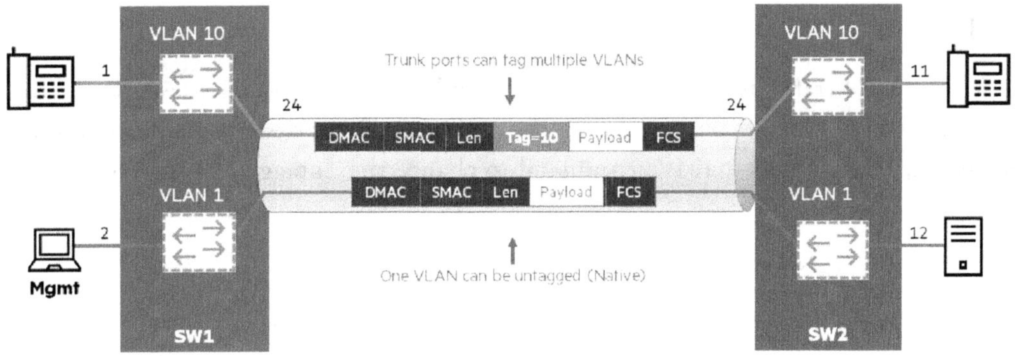

Figure 3-16: VLAN trunks carrying Ethernet frames

To avoid inconsistencies, HPE recommends you use the same native VLAN in both peers.

Configuring VLAN trunks

Let us now look at the steps to create VLAN trunks. In AOS-CX, use the `vlan trunk allowed` command to allow a VLAN-ID to traverse a trunk interface. In the following example, interface 1/1/24 is configured as a trunk link and only allows VLANs 1, 10, and 20. Other VLANs shall not traverse this link.

Configure trunk port

```
SW1(config)# interface 1/1/24
SW1(config-if)# vlan trunk allowed 1,10,20
```

This example focuses on trunk port configuration. It is assumed that the VLANs are already configured, and access ports are mapped to them.

To verify the trunk interface, you can use the `show vlan port <PORT-NUM>` command.

Validate trunk port

```
SW1# show vlan port 1/1/24
-------------------------------------------------------------
VLAN   Name             Mode                Mapping
-------------------------------------------------------------
1      DEFAULT_VLAN_1   native-untagged     port
10     Sales            trunk               port
20     Service          trunk               port
```

Native VLAN

By default, all VLAN traffic is tagged to traverse a trunk link except the native VLAN, which by default is VLAN 1. However, you can change the native VLAN as desired. As in the following example, use the `vlan trunk native` command to change the untagged VLAN, to 10 in this case.

Modify the native VLAN

```
SW1(config)# interface 1/1/24
SW1(config-if)# vlan trunk native 10
```

This trunk configuration must match on both switches, so be sure to apply this command to the appropriate port of the attached switch.

This is a good opportunity to remember earlier lessons. You can use the `show lldp neighbor-info` command to see which switch is connected to port 1/1/24 of the switch you are currently configuring.

Verify your efforts with the `show vlan port <PORT-NUM>` command.

```
SW1# show vlan port 1/1/24
-------------------------------------------------------------
VLAN    Name            Mode                Mapping
-------------------------------------------------------------
1       DEFAULT_VLAN_1  trunk               port
10      Sales           native-untagged     port
20      Service         trunk               port
```

Forwarding tables

Both forwarding tables and ARP tables play crucial roles in the efficient and accurate delivery of network frames within a LAN. So, now we will take a look first at MAC address tables and the ARP tables.

MAC address table

A MAC address table is where the switch stores information about other Ethernet interfaces to which it is connected on a network. Layer 2 switches use the MAC address table to make forwarding decisions. The switch builds this table automatically, based on the source MAC address of the frames that it receives from connected devices.

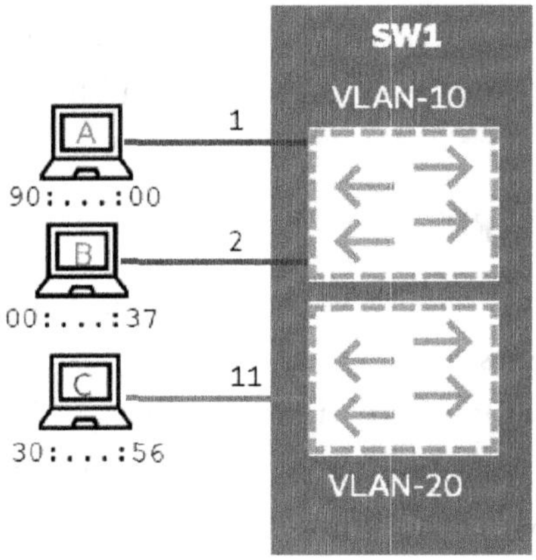

Figure 3-17: Generating a MAC address table

Consider a simple switch setup as shown in **Figure 3-17**. Suppose that you power up switch SW1. Its MAC address table is initially blank. Then, host A transmits a frame to host B. Switch 1 receives this frame with source MAC address = 90:...:00 on port 1, defined as being a member of VLAN 10.

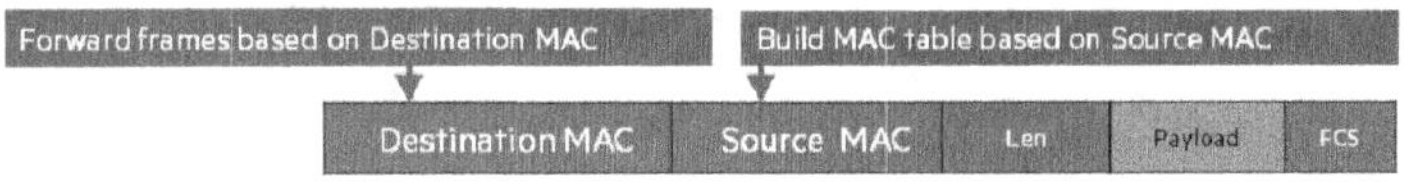

Thus, it adds the top entry to the MAC address table, as follows. It maps the MAC address to its VLAN and port.

```
SW1# show mac-address
MAC age-time            : 300 seconds
Number of MAC addresses : 3
MAC Address          VLAN     Type                Port
--------------------------------------------------------
90:20:c2:bc:ee:00    10       dynamic             1/1/1
00:0b:86:b4:eb:37    10       dynamic             1/1/2
30:8d:99:ef:84:56    20       dynamic             1/1/11
```

Suppose that, at this point, there are no other entries in the table. The switch does not yet know where to forward frames to destination 00:...:37 (MAC address of host B), so it floods this frame out all ports in VLAN 10. All VLAN 10 hosts receive this frame and see the destination MAC address. They say, "This is not for me." and they discard the frame. Except for host B, which says, "This is my MAC address, I will respond." So host B responds to host A with source MAC = 00:...:37 and destination MAC = 90:...:00.

Now the switch adds the second entry you see in the table: MAC 00:0b:86:b4:eb:37, in VLAN 10, connected to port 2. A similar process will happen with host C on port 11, with host C being in VLAN 20. So, switches automatically build the MAC table based on source MAC addresses and forward frames based on destination MAC addresses. By default, table entries are maintained for 300 seconds (five minutes). You can verify the MAC address table in AOS-CX using the `show mac-address` command.

Address resolution protocol (ARP)

ARP (Address Resolution Protocol) tables, on the other hand, are used by devices in a network to map IP addresses to MAC addresses. When a device wants to communicate with another device on the same network, it looks up the destination IP address in its ARP table to find the corresponding MAC address. If the MAC address is not present in the table, an ARP request is sent to the network to obtain the MAC address, and once received, it is added to the ARP table for future use.

Imagine that you are in a conference room, leading a discussion with people that you have just met. You do not know everyone's names yet. Someone enters the room and hands you a note that says, "Important message for Alvin Rogers." You do not know who Alvin is, so you say, "Excuse me, everyone. Who is Alvin Rogers?" Everyone hears your request, but only Alvin responds, "I'm Alvin." You can now associate the name Alvin to an actual person, seated at a particular chair in the room. This is very much like ARP, which maps Layer 3 IP addresses to Layer 2 MAC addresses.

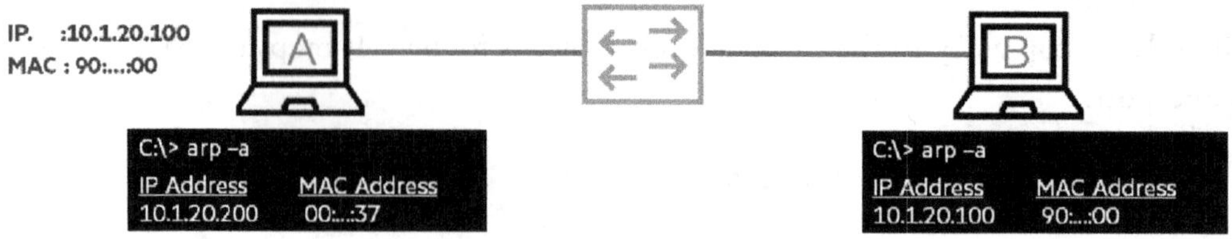

Figure 3-18: ARP scenario

In this scenario, host A knows that it must communicate with the host on its network at IP address 10.1.20.200. To do this, it must build an Ethernet frame, which requires its own source MAC address (90::00) and host B's destination MAC address, which it does not yet know. To learn it, host A broadcasts an ARP request—destination MAC address FF:FF:FF:FF:FF:FF. (In the IP header, the destination IP address is 255.255.255.255.)

Figure 3-19: ARP request Ethernet frame

Switches forward broadcasts out all ports in the same VLAN (except for the ingress port). Thus, all hosts in the VLAN receive this frame and ignore it ("I'm not 10.1.20.200"), except for host B, which responds with an ARP reply, a unicast to host A's MAC address, "Host A, 90:...:00, I am 10.1.20.100, and my MAC address is 00:...:37."

Figure 3-20: ARP reply Ethernet frame

Host A receives this reply and creates an entry in its ARP table, sometimes called an ARP cache. It maps 10.1.20.100 to 00:...:37. Thus, the next time host A must communicate with host B, it needs not use ARP. Also know that when host B received the ARP request from host A, it also learned host A's MAC address and IP address. Thus, host B creates an entry in its ARP table, mapping 10.1.20.100 to MAC address 90:...:00. At a Windows PC's command prompt, use the `arp -a` command to see which IP addresses have been resolved to a MAC address.

Frame delivery overview

Frame delivery in Layer 2 switch communication refers to the process of forwarding Ethernet frames between devices within a LAN based on their MAC addresses. Layer 2 switches use MAC address tables to make forwarding decisions, ensuring that data packets are delivered to the correct destination device while minimizing network congestion and optimizing data transmission efficiency.

In this section, you will learn how two devices in the same VLAN communicate across multiple switches. Consider the following scenario, where PC1 and Server-1 are each configured with an IP address.

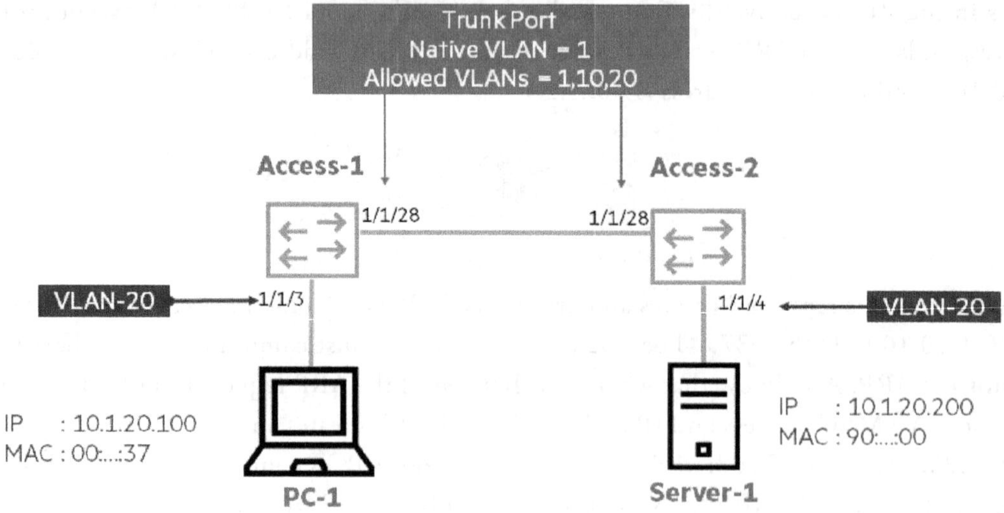

Figure 3-21: Frame delivery

Both switches are configured with three VLANs: 1, 10, and 20. The switch-to-switch links are configured as trunk ports allowing VLANs 1, 10, and 20, and the native VLAN = 1. Port 1/1/3 on the **Access-1** switch and port 1/1/4 on the **Access-2** switch are mapped to VLAN=20. Neither PC1 nor Server-1 have initiated any previous communications. The user on PC1 wants to establish an FTP session to download a file from Server-1. Let us now look at the various steps involved in this process.

In this guide, we cover Address Resolution Protocol (ARP), which is related to IPv4. In IPv6, ARP has been replaced by Neighbor Discovery (ND), a similar mechanism. It's important to note that IPv6 is beyond the scope of the current guide and certification.

Step 1

The user on PC-1 opens a browser and initiates an FTP session with Server-1: ftp://10.1.20.200. The FTP frame is as follows.

Layer 2 Header		Layer 3 Header		Layer 4 Header		Layer 5-7 Header	L2T
D-MAC	S-MAC	S-IP	D-IP	S-Port	D-Port	FTP GET	CRC
?	00:...:37	10.1.20.100	10.1.20.200	23561	21	Get **file.txt** from 10.1.20.200	

Step 2

PC-1 knows the IP address of Server-1; it does not know the MAC address associated to this IP. Therefore, an ARP process must be initiated. Also, notice that PC-1 has the rest of the information to build the Layer 2 frame, including Layer 3 to Layer 7.

Step 3

PC-1 must initiate an ARP request to identify the destination MAC address.

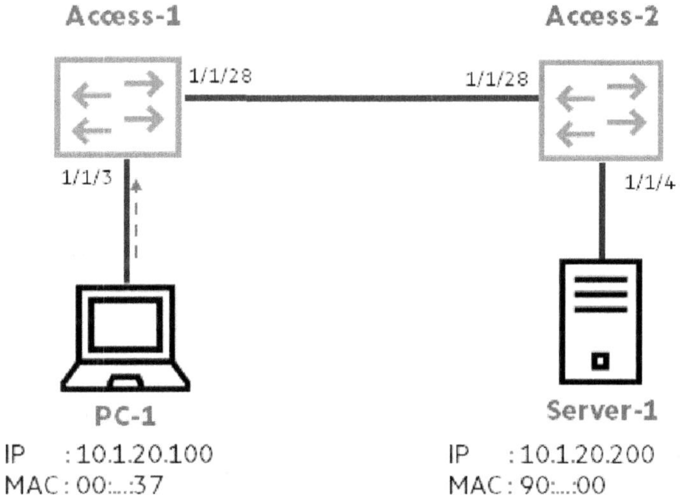

Figure 3-22: Initiating an ARP request

Step 4

PC-1 generates an ARP request. Notice that the ARP request has a Layer 2 broadcast destination MAC address (FF:FF:FF:FF:FF:FF) as follows.

Layer 2 Header		ARP Header				L2T
D-MAC	S-MAC	Sender MAC	Sender IP	Target MAC	Target IP	L2 Trailer
FF:...:FF	00:...:37	00:...:37	10.1.20.100	FF:...:FF	10.1.20.200	

Step 5

The frame is received by Access-1, which creates an entry in the MAC address table with the PC-1 MAC on port 1/1/3.

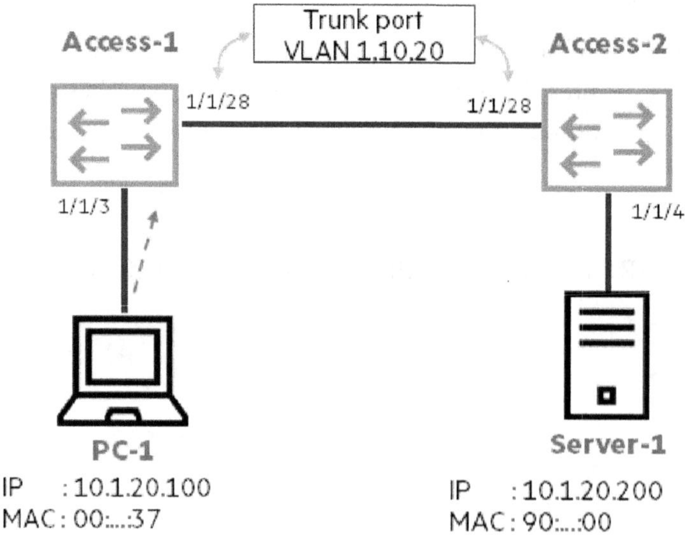

Figure 3-23: MAC address table entry created

Step 6

As the ARP request is a broadcast, the switch floods the frame out all the ports in VLAN 20, except for the port it was received from. In this example, the frame will be sent to the Access-2 switch via port 1/1/28. However, since port 1/1/28 is a trunk port and VLAN 20 is not the native VLAN, the frame will need to be tagged.

Step 7

Port 1/1/28 is configured as a trunk port, so the frame assigns an 802.1Q tag of VLAN-ID = 20 and forwards the frame.

Layer 2 Header			ARP Header				L2T
D-MAC	S-MAC	VLAN-ID	Sender MAC	Sender IP	Target MAC	Target IP	L2 Trailer
FF:...:FF	00:...:37	VLAN-ID = 20	00:...:37	10.1.20.100	FF:...:FF	10.1.20.200	

Step 8

The frame is received by Access-2. This switch will create an entry in the MAC table, with the MAC address of PC-1 and port 1/1/28.

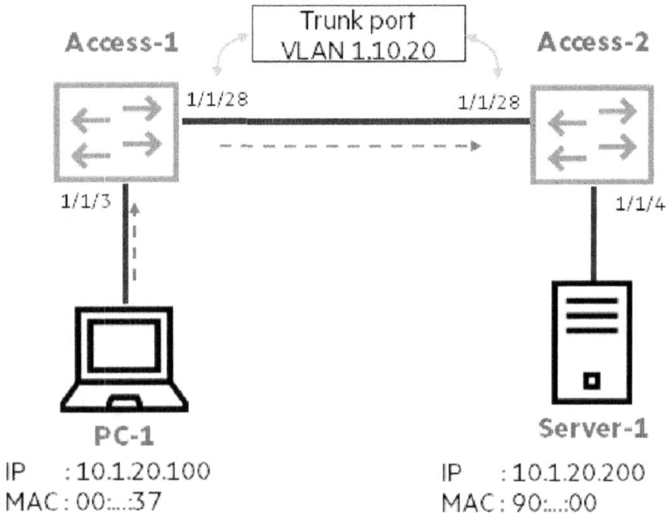

Figure 3-24: 802.1Q tagging

Step 9

Access-2 forwards the frame to all the ports mapped to VLAN 20. In this case, the frame is sent out interface 1/1/4. This interface is not a trunk port, which implies that the frame will not include an 802.1Q tag with a VLAN-ID field.

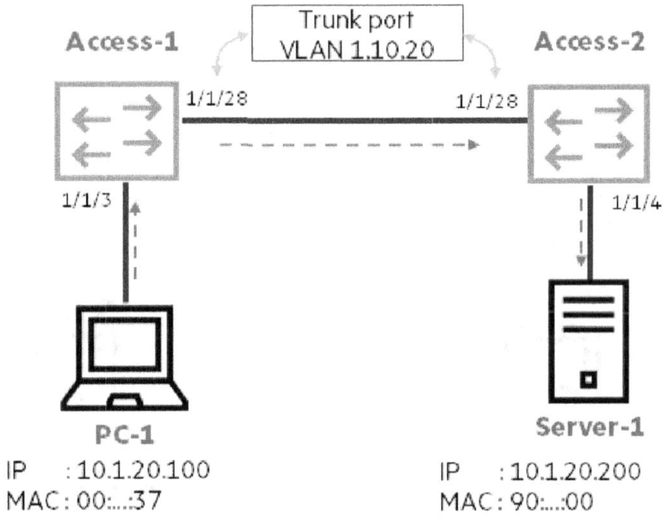

Figure 3-25: Forward untagged frame

Step 10

Server-1 adds the PC-1 IP address to its ARP table.

Layer 2 Header		ARP Header				L2T
D-MAC	S-MAC	Sender MAC	Sender IP	Target MAC	Target IP	L2 Trailer
FF:...:FF	00:...:37	00:...:37	10.1.20.100	FF:...:FF	10.1.20.200	

Step 11

Server-1 receives the ARP request and generates an ARP reply. Server-1 knows the source MAC, so this is a unicast frame.

Layer 2 Header		ARP Header				L2T
D-MAC	S-MAC	Sender MAC	Sender IP	Target MAC	Target IP	L2 Trailer
90:...:00	00:...:37	00:...:37	10.1.20.100	90:...:00	10.1.20.200	

Step 12

Access-2 processes the ARP reply message. The switch learns the Server-1 MAC address and adds the entry to the MAC address table.

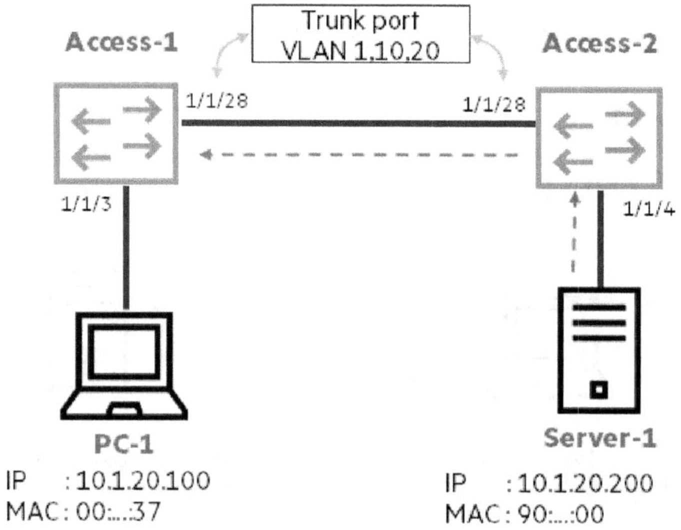

Figure 3-26: ARP reply

Step 13

Access-2 sends the ARP reply to Access-1. This frame is tagged, so it includes the VLAN-ID of 20 and the frame is sent out from interface 1/1/28.

Step 14

Access-1 forwards the ARP reply to PC-1 via port 1/1/3. This port is not a trunk, so no VLAN tag is used.

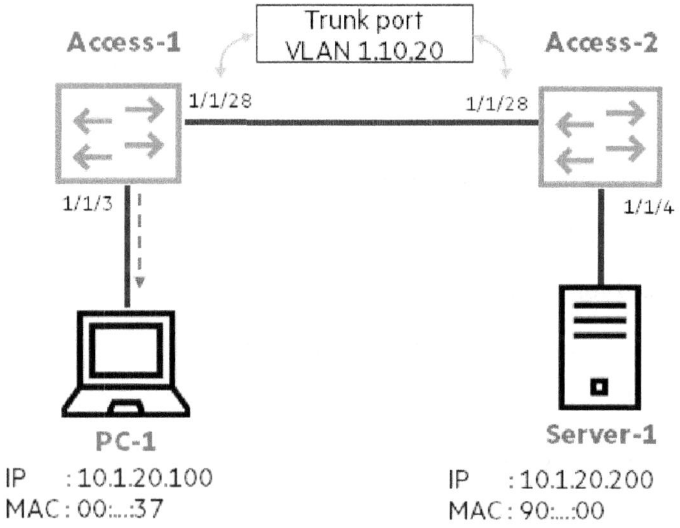

Figure 3-27: PC receives an ARP reply

Step 15

The switch creates entries for Server-1 in the MAC and ARP tables.

Step 16

Now, PC-1 knows Server-1's MAC address and adds it to its ARP table. PC-1 can build the Ethernet frame and initiate the FTP communication.

Learning check

Check your understanding of this material by answering the following questions.

Question 1

Which of the options below accurately describes the difference between collision domains and broadcast domains? (Select two)

- ☐ Collision domains relate to Layer 2 processes, while broadcast domains are a Layer 3 concept.
- ☐ Collisions are quite common in modern network switches.
- ☐ A routing device defines the edge of a broadcast domain.
- ☐ A switch port defines the edge of a collision domain.
- ☐ Broadcast domains are mainly a problem when you use a hub device.

Question 2

What are the benefits of creating VLANs? (Select two)

- ☐ Simplify the network deployment
- ☐ Minimize the need for routing
- ☐ Smaller broadcast domains can improve performance
- ☐ MAC address tables are smaller
- ☐ Separate VLANs mitigate risk

Question 3

What is true from the following lines of configuration?

```
SW1(config)# interface 1/1/24
SW1(config-if)# vlan trunk allow 1,20
```

- ○ VLAN-20 is the native VLAN.
- ○ VLAN-10 is permitted in the trunk port.
- ○ There is no native VLAN.
- ○ Native VLAN is VLAN-1.

4 Spanning tree protocol

Welcome to the Spanning Tree Protocol (STP) chapter, where we are going to explore the fundamental concepts and mechanisms behind an important protocol in Ethernet networks. STP is essential for preventing network loops and ensuring network redundancy by intelligently managing redundant links. We will be learning about the purpose, operation, and configuration of STP, as well as its variants and advanced features, to equip you with the knowledge and skills needed to optimize stability and performance.

OBJECTIVES

After completing this chapter, you should be able to:

- ✓ Identify the purpose and operation of the Spanning Tree Protocol.
- ✓ Mitigate the risks associated with redundant network links and Layer 2 loops.
- ✓ Configure Rapid Spanning Tree Protocol (RSTP).
- ✓ Integrate RSTP concepts such as elements, operation, ports, links, proposals, and agreements.
- ✓ Analyze Multiple Spanning Tree Protocol (MSTP) to evaluate its benefits, operation, and configuration requirements.

Purpose

The purpose of Spanning Tree Protocol (STP) is to maintain a redundant loop-free topology by dynamically identifying and disabling redundant paths.

STP operates by electing a root bridge, determining the shortest path to reach each network segment, and blocking redundant links to ensure a loop-free topology. By continuously monitoring the network topology and adapting to changes, STP ensures network stability and prevents broadcast storms caused by loops.

Types of STP

There are several types of Spanning Tree Protocols and their variants, each designed to address specific network requirements and challenges.

Figure 4-1: STP, RSTP, MSTP types

Spanning Tree Protocol (STP)

The original IEEE 802.1D standard protocol used to prevent loops in Ethernet by disabling redundant paths and creating a loop-free topology. It operates based on the Spanning Tree Algorithm (STA) and provides basic loop prevention mechanisms.

Rapid Spanning Tree Protocol (RSTP)

An enhancement to STP defined by IEEE 802.1w, RSTP improves convergence times by introducing new port roles states and reducing the time required to transition between them. It allows for faster network convergence and better utilization.

Multiple Spanning Tree Protocol (MSTP)

Defined by IEEE 802.1s, MSTP allows for the creation of Multiple Spanning Tree Instances (MSTIs) to map multiple VLANs to a single spanning tree topology. This offers better scalability and flexibility than STP and RSTP, particularly in networks with multiple VLANs.

Redundant network

We are going to begin with the critical concepts surrounding network redundancy and potential pitfalls associated with Layer 2 loops in networking environments. By understanding the importance of redundancy for network resilience and the risks posed by Layer 2 loops, network administrators can effectively design and manage robust network architectures.

Redundant network links

We are going to briefly talk about the redundant network and loops. This discussion sets the foundation for the reason behind the Spanning Tree Protocol.

Now, network communications are most often mission critical, meaning Layer 2 switch or link failures cannot be tolerated. For example, if you have a single path to connect two access switches using Core-1, this design has a single point of failure. If Core-1 fails, the network is down. Hosts A and B cannot communicate.

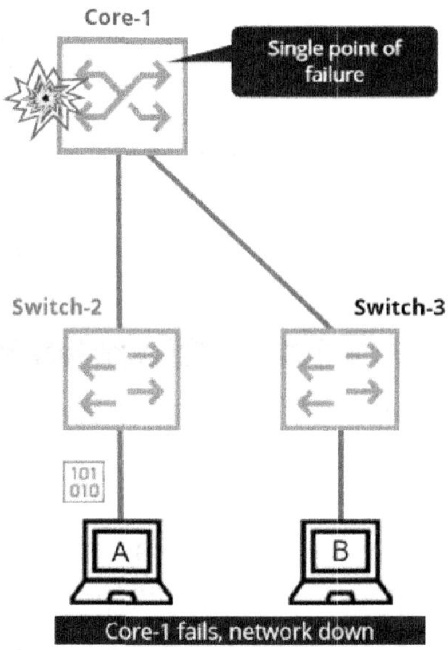

Figure 4-2: Single point of failure

One common way to mitigate this is by adding a redundant core switch. If Core-1 fails, the network remains viable. Hosts A and B can still communicate over the redundant link using Core-2.

While this redundant link mitigates a single point of failure issue, it creates a new challenge.

Now, connecting Layer 2 switches with redundant links creates Layer 2 loops. A loop is even created if you connect an Ethernet cable from one port on a switch to another port on that same switch.

If not properly handled, these loops cause serious problems that can effectively disable your network. The following items reflect the three main problems that result from redundant layers.

Layer 2 loop

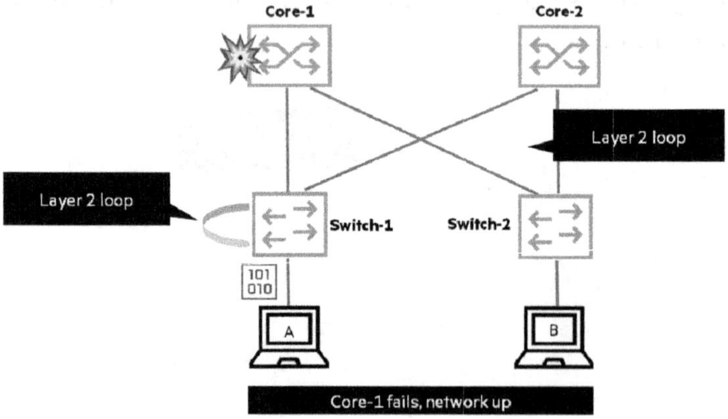

Figure 4-3: Redundant Core-2

Broadcast storm

Switches react to broadcast messages, or a unicast message with an unknown address. The switch will forward that frame out of every port, except the one that received it.

Suppose that host A sends a broadcast frame. Switch-1 receives this frame and floods it out all interfaces (Copy 1). Core-1 and Core-2 receive this and flood it out their connections to each other (Copy 2). Core-1 and Core-2 receive the second copy and flood it out their other ports (Copy 3). Switch-1 receives the third copy, and the cycle continues. These frames circle around the network continuously.

As a natural part of network and endpoint operation, nearly all devices send broadcast frames. Every broadcast from every device circulates around the network forever. Soon, all available bandwidth and CPU cycles are used up processing broadcasts. No resources are available to process normal data communication frames; the network is effectively unusable. This is a broadcast storm.

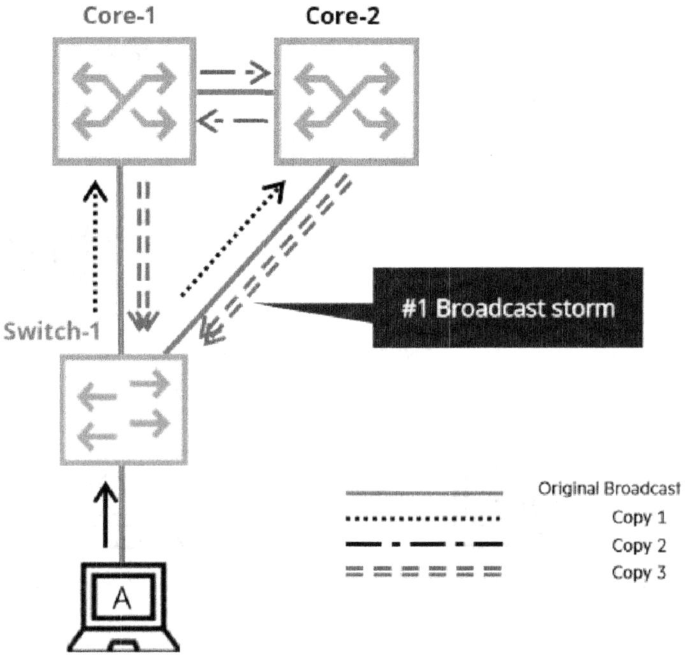

Figure 4-4: Broadcast storm topology

Unstable MAC address table

Switch 1 sends a broadcast message to Core-1 and Switch 3. Core-1 does exactly what switches are meant to do, and it learns the MAC address of the message. Then what happens when Core-1 gets the same broadcast message from Switch 3 on another port? It learns the MAC address, and it is thinking "Hey, I just got this MAC address on 03 and now I'm being told it is on 05." Core-1 will update the MAC address table to 05.

The same message will keep appearing in all ports. How often is this MAC address table getting updated? And since the broadcast storm is happening on all switches, they all will be updating their MAC address tables continuously. This creates the second problem with switching loops, which is unstable MAC address tables.

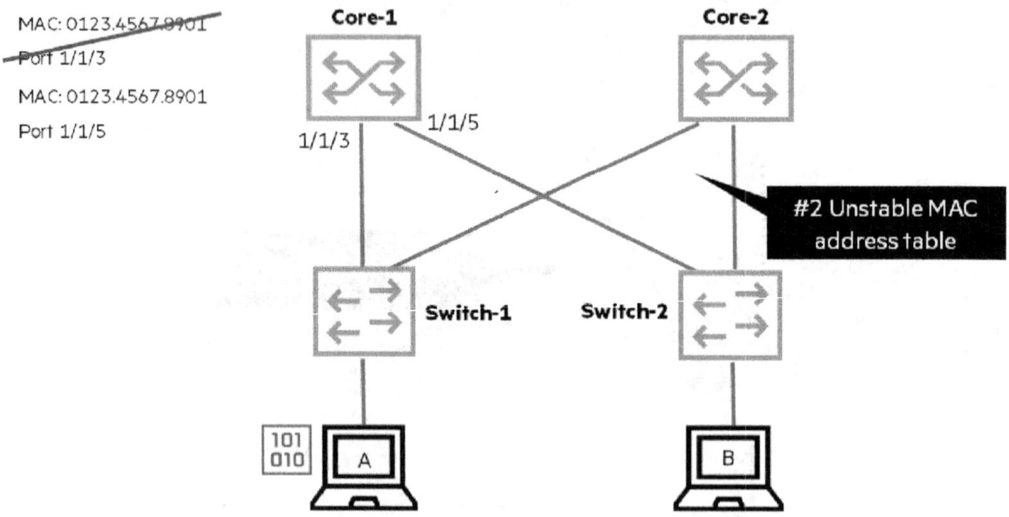

Figure 4-5: Unstable MAC address table topology example

Duplicate frames

Host B sends data to host A. If Switch 3 does not know the address of host A's MAC address, it will flood the frame of all other ports. Core-1 received it and forwards it out of a port to Switch-1, which delivers the frame to host A. Core-2 now forwards the packet to Switch-1. Switch-1 does not recognize it is a duplicate frame and sends the packet to host A again. Now we have duplicate frames being sent all over the network.

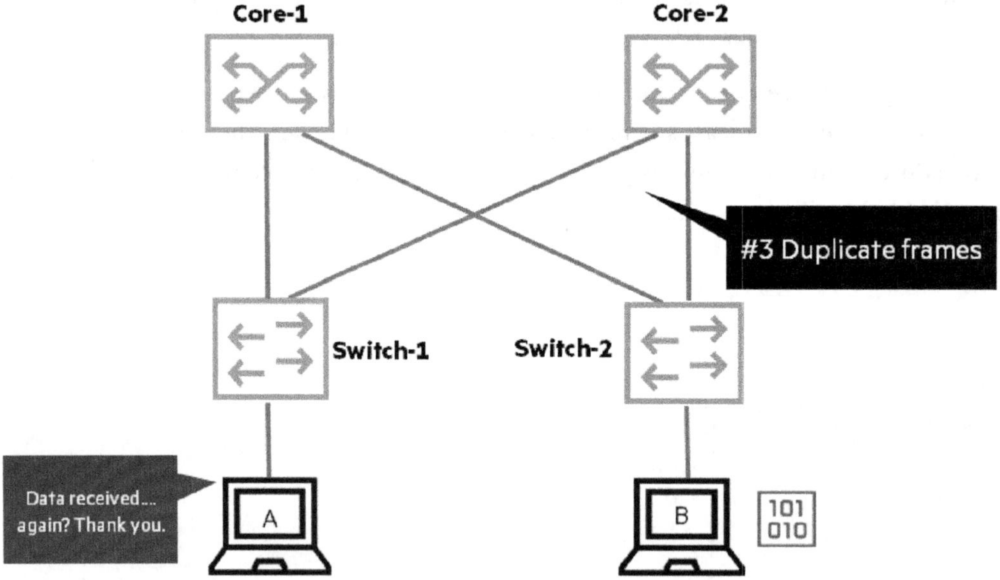

Figure 4-6: Duplicate frames topology

Spanning Tree Protocol

STP remains essential for preventing loops and ensuring a loop-free topology in Ethernet networks, complementing the benefits provided by Link Aggregation Groups. Additionally, newer versions of STP, such as Rapid Spanning Tree Protocol and Multiple Spanning Tree Protocol, offer improved convergence times and scalability. We are going to start by going over the original Spanning Tree Protocol, which is the foundation for this. Then we will jump into a more updated Rapid Spanning Tree Protocol.

STP is a protocol after all, and it has a specific process to follow to determine which port to block. Once you can understand this process in detail, STP starts to become much clearer and easier to understand.

This is a high-level overview of how STP functions:

1. Select a route bridge, and this is your CEO of switches.
2. STP places all root bridge interfaces into a forwarding state.
3. Each non-root switch selects a root port, and this is the best route to the port.
4. The remaining links will choose whether or not they become designated ports.
5. All other ports are placed into a blocking state.

While that seems easy, how the switches choose which port to block is more complicated.

STP overview

Why is it called Spanning Tree Protocol? If you recall, we have three layers: the core layer, the aggregation layer, and the access layer. The access layer connects to the edge ports. If you think of a tree, you have the roots, the trunk and branches, and the leaves. The edge ports are like the leaves of the tree.

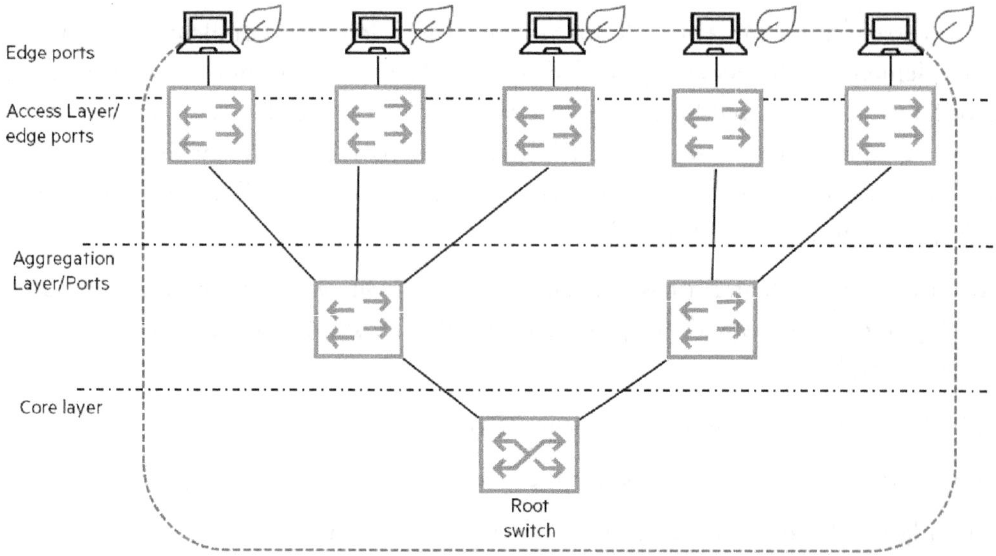

Figure 4-7: Spanning tree protocol representation

The IEEE 802.1D standard version of STP was developed to build and maintain redundant, loop-free networks. With STP, you eliminate single points of failure, while avoiding loops and MAC table instability. STP creates a loop-free topology by automatically disabling redundant links.

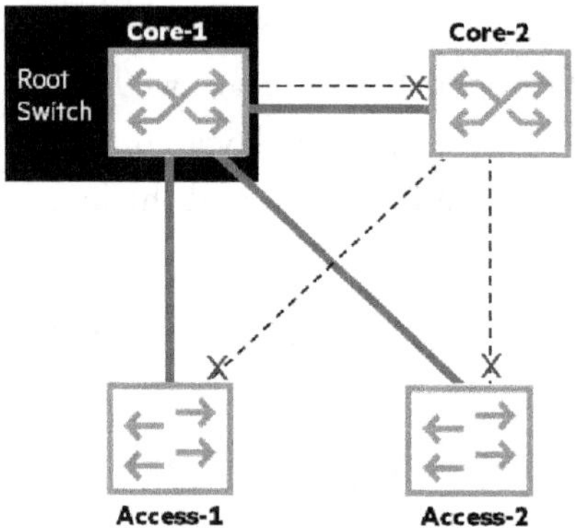

Figure 4-8: Layer 2 loop-free topology

This is a highly redundant network, for resiliency and fault tolerance. These redundant links could cause loops and their associated issues.

Once STP is engaged, certain redundant links are automatically disabled; there are no loops, so there are no problems with broadcast storms or MAC address table instability. However, these disabled links remain available to provide redundancy as needed.

To accomplish its goal, one switch in the spanning tree domain is elected as a root bridge or root switch. Many trees grow out of the ground, from its roots, as a single trunk and then branch out from there. Likewise, the root bridge is the reference from which the spanning tree grows. All other switches are non-root bridges (switches), sometimes knows as designated switches. A loop-free path "grows" from this root switch out to all non-designated (non-root) switches.

What is now called a "switch" is a very fast, multiple-port version of what used to be called a "bridge" (many years ago). When discussing STP, you will often see the term "bridge." In this context, "bridge" and "switch" can be thought of as the same thing. When you see "bridge," know that the discussion is about a switch.

Reconvergence

The STP algorithm runs on all switches, and redundant links have been disabled to avoid loops.

However, if a failure occurs, STP converges on a new topology of active links, which are used to forward frames. A loop-free topology is always maintained.

Reconvergence is what happens when a failure occurs, and loop-free topology is always maintained.

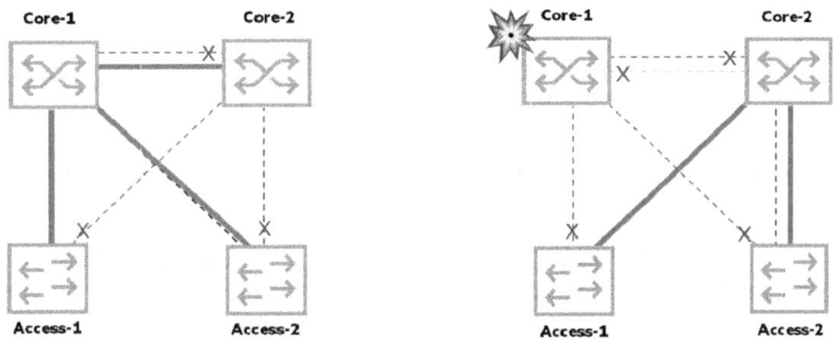

Figure 4-9: Reconvergence to keep a loop-free topology

The word "converge" can mean something like "to meet at some point." Convergence, in the context of networking, means that the devices all come to an agreement about a new network topology. All devices converge on a new set of active paths to be used for frame forwarding.

STP IEEE protocol evolution

In the original standard, failure detection is based on timers. With the IEEE 802.1D default, the root switch (and only the root switch) originates a "hello" packet every two seconds. These hello packets are forwarded on out to the rest of the switches in the domain. If some switch downstream (farther away from the root switch) ceases to receive hello packets for 20 seconds (the default Max Age timer), an outage is assumed, and all switches begin to converge on a new topology. The duration of this process is defined by the Forward Delay timer, which is 15 seconds by default. This protocol is considered obsolete, and its use is no longer recommended in modern networks.

Rapid Spanning Tree Protocol (RSTP) was developed in 1998 to speed convergence. Instead of only the root switch originating hello packets, ALL switches originate them. This means that RSTP now has a true keepalive mechanism that can respond in seconds (or less). The need for old and slow Max Age and Forward Delay timers is eliminated. You will soon learn the details of this new convergence process.

> RSTP is backward compatible with the original standard 802.1D. However, to maintain this capability, some benefits of RSTP are lost.

802.1s or Multiple Spanning Tree was developed to improve the performance of the protocol for implementing multiple loop-free topologies or instances to load balance the traffic across all links. This protocol helps to create optimal paths.

MSTP runs on AOS-CX switches by default. However, if no setting is configured, then this protocol behaves like RSTP. You will learn why this happens later in this training.

Table 4-1: STP, RSTP, and MSTP comparison

STP	RSTP	MSTP
802.1D (pre-2004)	**802.1w** merged into **802.1D 2004**	**802.1s** merged into **802.1Q 2005**
■ Original protocol ■ Slow convergence	■ Faster convergence ■ Costs that fit with modern port speeds	RSTP behavior between multiple instances
■ Obsolete ■ Not recommended	Default operation between AOS-CX switches	Default mode for AOS-CX switches

Rapid Spanning Tree Protocol

Rapid Spanning Tree Protocol (RSTP) was created to address some of the limitations and shortcomings of the original STP, including:

Faster convergence

One of the primary reasons for developing RSTP was to improve network convergence times compared to STP. RSTP introduces new port states and mechanisms to reduce the time required for the network to transition to a stable state after topology changes, resulting in faster convergence and reduced downtime.

Efficient use of redundant links

RSTP optimizes the use of redundant links in the network by allowing for rapid transitioning between different port roles (example, blocking, forwarding) in response to topology changes. This allows for better utilization of available bandwidth and improved network efficiency compared to STP.

Simplified configuration

RSTP simplifies the configuration and management of Spanning Tree Protocols by reducing the number of port states and introducing backward compatibility with transitional STP. This makes it easier for network administrators to deploy and maintain RSTP-based networks compared to STP.

Support for larger networks

RSTP is better suited for larger and more complex networks compared to STP, thanks to its faster convergence times and improved scalability. It allows for quicker adaptation to changes in network topology, making it suitable for modern network environments with dynamic workloads and frequent changes.

Overall, RSTP was developed to address the evolving needs of modern network infrastructures, providing faster convergence, improved efficiency, simplified configuration, and better scalability.

RSTP has four main sections:

1. RSTP elements, like Bridge ID, Bridge Protocol Data Unit (BPDU), port state, port cost, and port roles.
2. Operation, which builds the loop-free topology.
3. Edge ports and link types, which help optimize the spanning tree topology and improve network performance.

4. Topology change mechanism: When a topology change occurs, such as a link failure or recovery, RSTP uses a specific mechanism to minimize the impact on network performance and stability.

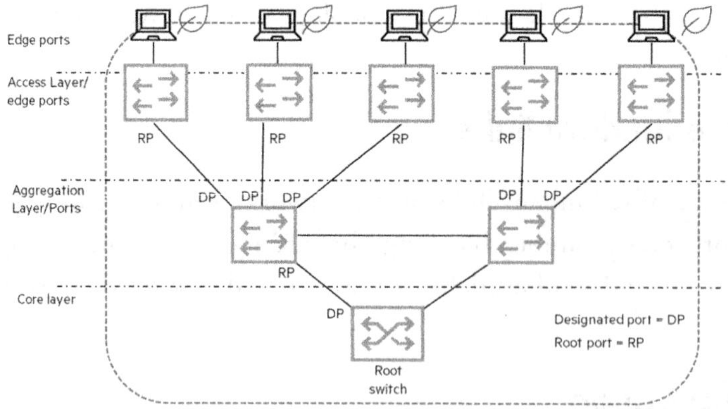

Figure 4-10: RSTP representation with designated ports and root ports

RSTP elements overview

These are the key components that the rapid spanning tree algorithm uses to decide which ports to enable or not for traffic forwarding.

They includes the following:

- Bridge Identifier
- Bridge Protocol Data Unit
- Rapid Spanning Tree Port States
- Rapid Spanning Tree Port Cost
- Rapid Spanning Tree Port Roles

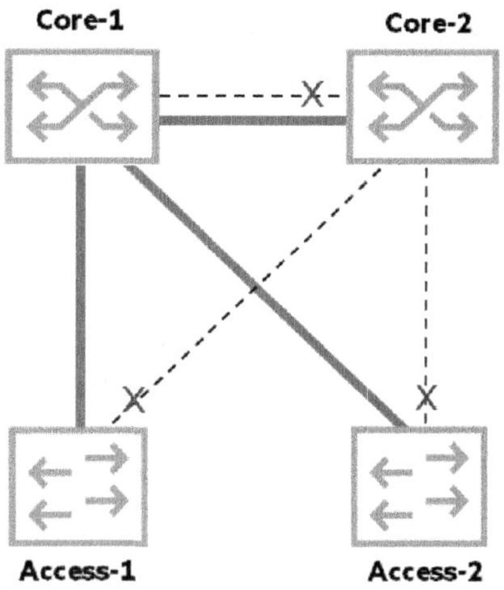

Figure 4-11: RSTP elements overview

Bridge identifer

In RSTP, the bridge identifier—also known as the Bridge ID or BID—is a unique identifier assigned to each switch participating in the spanning tree topology. The bridge identifier consists of two components:

1. **Bridge priority:** A two-byte priority (ranging from 0 to 65,535) that determines the priority of the switch in the spanning tree topology. Switches with lower bridge priority values are preferred as root bridges.
2. **MAC address:** A six-byte MAC address of the switch, which serves as a unique identifier for the switch within the network.

The priority defaults to 32,768, so by default, all switches have the same priority. However, each switch has a unique MAC address, and so each BID will be unique.

The bridge identifier is used during the spanning tree election process to determine the root bridge and to calculate the shortest path to the root bridge for each switch in the network. The switch with the lowest bridge identifier (a combination of the lowest bridge priority and the MAC address) is elected as the root bridge, and all other switches will adjust their spanning tree paths accordingly.

By using bridge identifiers, RSTP ensures that switches can accurately determine the topology of the network and establish the most efficient spanning tree configuration for optimal traffic forwarding and network stability.

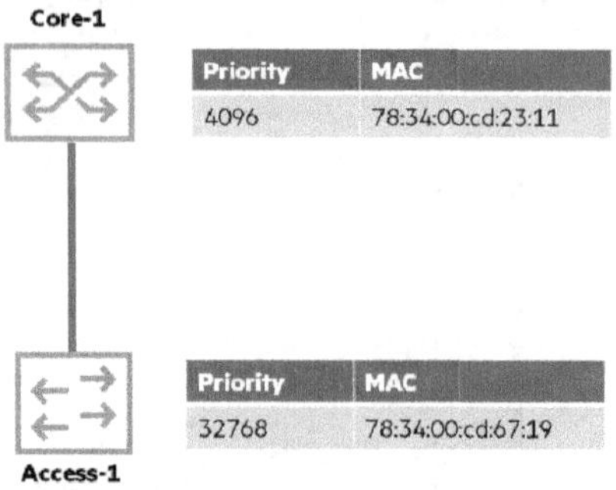

Figure 4-12: BID topology

Bridge protocol data unit

A Bridge Protocol Data Unit, known as BPDU, is a message format used by switches running RSTP to exchange information and negotiate parameters for establishing and maintaining the spanning tree topology.

RSTP BPDUs carry essential information, such as:

- Bridge ID
- Port role and state
- Path cost
- Hello time and Forward Delay
- Topology Change Notification (TCN)

In the original 802.1D standard, BPDUs were generated only by the root switch and then had to "trickle down" to other switches. This led to the need for the slow Max Age and Forward Delay timers.

With RSTP, all switches originate BPDUs with their current information every two seconds, the default hello-time period. Thus, if a port stops receiving BPDUs for three consecutive hello timers, the switch quickly knows that it has lost connectivity to its neighbor. It ages out the protocol information and begins to converge.

Because each switch originates BPDUs, this becomes a true keepalive mechanism. Failure detection will take no longer than six seconds.

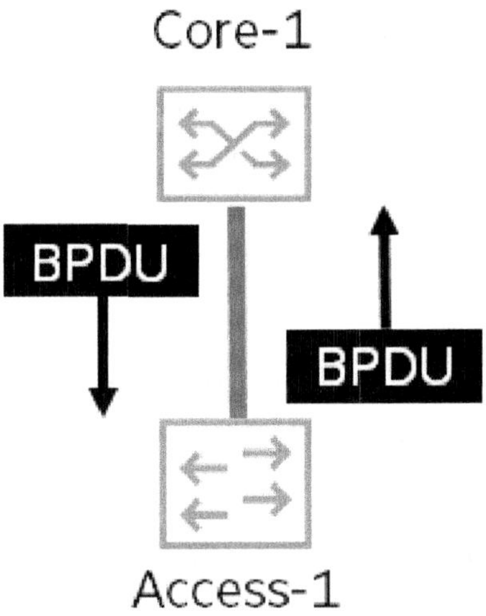

Figure 4-13: BPDU topology

Port states

During original tree establishment and any ensuing convergence, switches must decide which ports will forward data and which ports must be disabled to prevent Layer 2 loops. Spanning tree uses port states to transit from a blocking port to a forwarding port. The following image summarizes the port states and their specific tasks. This lists the states that were used in 802.1D and compares them against the new port states in 802.1w.

To transition from Blocking to Forwarding, the original 802.1D standard takes 30 seconds—15 seconds are spent in a Listening state and 15 seconds in the Learning state.

One of the reasons RSTP is more efficient than 802.1D is because the port transitions quickly from Discarding to Learning to Forwarding.

STP (802.1D) port state	RSTP (802.1w) port state	Port included in active topology / Port learning MAC addresses?	BPDU receiving	BPDU sending
Disabled	Discarding	X	X	X
Blocking		X	✓	X
Listening -15 seconds		X	✓	✓
Learning- 15 seconds	Learning – 2 seconds	✓	✓	✓
Forwarding	Forwarding	✓	✓	✓

Figure 4-14: STP port state compared to RSTP port state

The Blocking and Listening states are also deprecated in MSTP.

In RSTP, the Learning state is a transitory state and is only used during a reconvergence of the protocol when a change happens in the topology. Stable states will be the Discarding or Forwarding states.

Path cost

RSTP may have several possible paths to get from the root switch to some non-root switch. It chooses the best path based on cost, which is based on link speed. **Table 4-3** shows the AOS-CX default path costs.

Consider the example shown in **Figure 4-15**. Intuitively, you might think the root switch's best path to Access-1 is the direct path.

However, this is a 1 Gbps link, which has a cost of 20,000.

The indirect path to Access-1 via Core-2 uses two 10 Gbps links, which only have a cost of 2000 each. 2000 + 2000 = 4000—far less than 20,000. (Strictly speaking, the root switch will add its cost to the calculation; in this case its cost is equal to 0.)

Thus, the indirect path is the preferred, best path, and the redundant path is disabled to avoid loops.

BPDUs advertise the path cost.

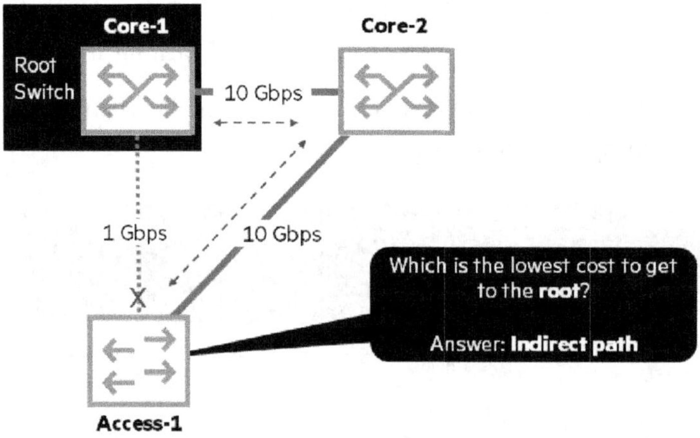

Figure 4-15: Path cost topology

Table 4-2: Direct and indirect path cost

Direct path	Indirect path
0 + 20,000 = 20,000	0 + 2,00 + 2,000 = 4,000

Table 4-3: Link speed and RSTP cost

Link speed	RSTP port cost in AOS-CX
100 Mpps	200,000
1 Gbps	20,000
10 Gbps	2000
100 Gbps	200
1 Tbps	20

Port roles

Switches that use 802.1w RSTP do not have a complete view of the topology. They build and maintain the loop-free topology by exchanging BPDUs, which indicate how close they are to the root switch. BPDUs help switches to calculate the correct port role for each of their ports.

All ports on the root switch are designated ports. The root switch is like the "boss" of the domain. It does not block its ports; only non-designated ports must worry about that.

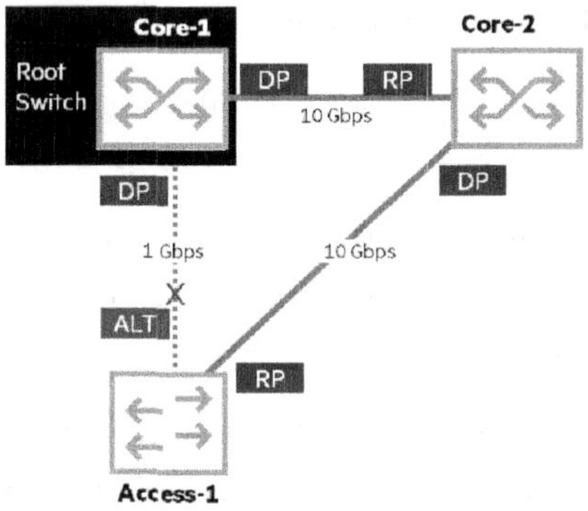

Figure 4-16: Port roles (designated, root, and alternate)

There are no loops in the topology, so all ports are forwarding. Looking at the non-root switches, what is the difference between the root port and designated port? They both forward frames. The root port is simply the port a switch uses to reach the root: the best path toward the root switch. The designated port is designated to accept traffic from downstream root ports.

Designated port

- Closest to the root switch
- All root switch ports are designated
- Port state = **Forwarding**

Root port

- Another switch in the link is closer to the root
- Best path to the root switch (for a non-root switch)
- Port state = **Forwarding**

Alternate port

- Not the closest to the root switch
- No loop to itself
- Becomes the root port if the active port fails
- Port state = **Discarding**

Backup ports

A backup port is a type of port role that provides backup connectivity to a segment in the event of a failure on the designated port. Backup ports are used to restore connectivity quickly in case the primary path becomes unavailable, ensuring rapid convergence and minimal network downtime.

Key characteristics of RSTP backup ports include:

- **Role:** A backup port is a non-designated port that is in a Discarding state. It serves as a standby port that becomes active if the designated port on the same segment fails.
- **Functionality:** Backup ports monitor the designated port on the same segment for any failures. If the designated port fails, the backup port transitions to the forwarding state to maintain connectivity to the segment.
- **Priority:** Backup ports have a lower priority than designated ports, meaning they are less preferred for forwarding traffic under normal conditions. However, they are ready to take over forwarding duties if needed.
- **Path cost:** Like other port roles in RSTP, backup ports calculate the path cost to the root bridge. This information is used to determine the shortest path to the root bridge and to select the best forwarding path in the network.

Listed are the characteristics of a backup port:

- Not the closest to the root switch
- Port state = discarding
- Creates a loop to itself
- Becomes the designated port if the active port fails

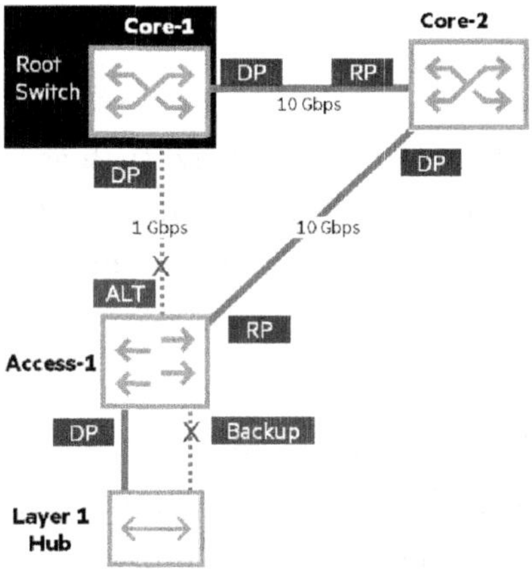

Figure 4-17: Backup port topology

A loop could be caused when a Layer 1 hub is placed on the topology. As an example, take into consideration **Figure 4-17**. In this situation, Access-1 has two connections that are in a loop situation because a hub was introduced to the topology. If RTSP was running on Access-1, neither of the ports connected to the hub would become an alternate port.

A backup port is considered when a hub is connected to the network.

Introducing Layer 1 hubs into the topology should be avoided.

RSTP operations overview

RSTP operation refers to the process and behavior of RSTP-enabled switches within a network. This includes how switches exchange information, calculate spanning tree paths, and forward traffic to ensure a loop-free topology and optimal network performance.

The RSTP algorithm converges on a loop-free topology by performing the following steps:

1. Elects root switch and backup root switch.
2. Selects the root port on all non-root switches or non-designated bridges.
3. Selects the designated port on each switch-to-switch link.
4. Other ports move to a blocking state to prevent loops.

Key aspects of RSTP operation include:

Port roles and states

RSTP defines different port roles and port states to determine how switches forward traffic and participate in the spanning tree topology. These roles and states are dynamically adjusted based on changes in the network topology.

Topology changes

RSTP rapidly detects changes in the network topology, such as link failures or recoveries, and initiates a rapid convergence process to update the spanning tree topology accordingly. This ensures that switches quickly adapt to changes and maintain network connectivity without disruption.

Path calculation

RSTP calculates the shortest path to the root bridge within the spanning tree topology, considering the cost associated with each link. This calculation ensures that traffic is forwarded along the most efficient path, minimizing latency and maximizing network performance.

BPDUs

RSTP-enabled switches exchange BPDUs to communicate and negotiate spanning tree parameters, such as root bridge selection, port roles, and path costs. These BPDUs allow switches to establish and maintain a consistent spanning tree topology across the network.

Rapid convergence

RSTP aims to achieve rapid convergence by minimizing the time it takes for switches to transition ports between different states in response to topology changes. This rapid convergence ensures that network downtime is minimized, and traffic continues to flow smoothly in dynamic network environments.

Root switch election

For both the 802.1D and 802.1w standards, root switch election is based on the Bridge ID (BID)—the lowest value wins. Recall that the two-part BID is composed of a priority (which defaults to 32,768 for all switches) and a globally unique MAC address.

You can rely on this default, rather random behavior, but it is not recommended. Some small, low-powered switch on the edge of your network might win the election. This makes for a less stable tree, with sub-optimal paths and poor resiliency. Consider the topology shown in **Figure 4-18**. If the priority is set to the default, then Access-2 will become the root switch as it has the lowest MAC address, and Access-2 will become the backup root switch.

You want to ensure that one of the high-powered, more centrally located core switches wins the election. You will have a more robust, resilient, and optimally pathed tree structure. To do this, you simply lower the priority value of the preferred switches, lower than the 32,768 default. In this example, Core-1 becomes the root switch simply by setting its priority to 4096. However, if Core-1 fails, then Access-2 would become the root switch, which would not be optimal. Therefore, it is important that Core-2 becomes the second-best option. In this case, Core-2 is set up with a priority of 8192.

Recall that the priority value must be configured in increments of 4096. You might lower the value of Core-1 to 4096 and the value of Core-2 to 8192.

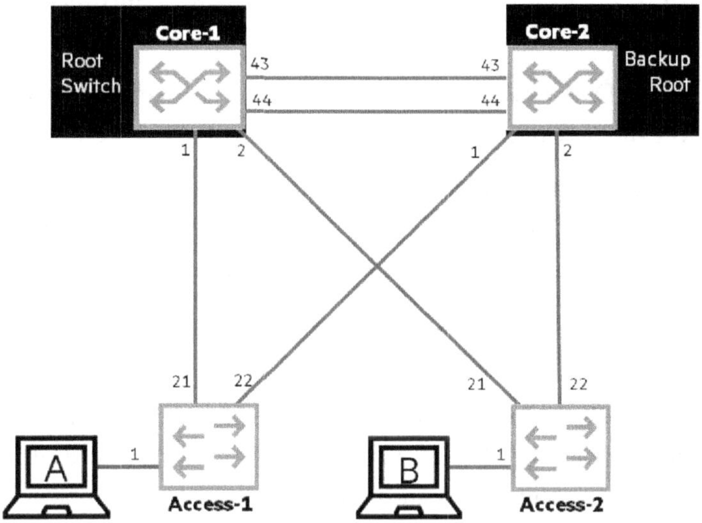

Figure 4-18: Root switch election

Table 4-4: Device, priority, and MAC root switch selection

Device	Priority	MAC
Core-1	4096	00:....:04
Core-2	8192	00:....:03
Access-1	32,768	00:....:02
Access-2	32,768	00:....:01

Select root ports in non-root switches

Non-root switches, or non-designated bridges, must select the best path to the root switch, by selecting its root port—the port connected to this best path. Root port selection criteria are as follows:

- Lowest root bridge ID
- Lowest path cost to the root bridge
- Lowest sender bridge ID
- Lowest port priority
- Lowest port ID

Core-2 analysis

This device will receive BPDUs on ports 1, 2, 43, and 44. Core-2 must decide which port is the best one.

- Lowest root bridge ID. All devices agree that the root bridge is Core-1; this criterion does not help to select the best path.
- Lowest path cost to the root bridge. Assuming that all links are the same speed, then the indirect paths on ports 1 and 2 are discarded simply because the cost is higher than the direct paths. Since ports 43 and 44 have the same cost, the next criterion must be used.
- Lowest sender bridge ID. BDPUs received on ports 43 and 44 are generated by the same device (Core-1); this criterion does not help to break the tie.
- Lowest port priority. The sender (Core-1) will include a port priority on the advertisement. The port priority is set as a multiplier value of 0-15. The value is multiplied by 16, which generates a priority in a range between 0-240. The default multiplier is 8, except LAG connections default to a multiplier of 4. Since the AOS-CX default value is 128 (8 x 16), a bridge will consider a lower number the winner. In this case, assuming that default values are set, this criterion does not help to break the tie.
- Lowest port ID. The final decision is taken on the port ID, where the lowest value is the winner. Core-1's lowest port number is 43, so the port on Core-2 that is connected to this port becomes the root port (in this case, Core-2's port 43).

Access-1 analysis

This device will receive BPDUs on ports 21 and 22. Access-1 must decide which port is the best one.

- Lowest root bridge ID. All devices agree that the root bridge is Core-1; this criterion does not help to select the best path.

- Lowest path cost to the root bridge. Assuming that all links are the same speed, then the indirect path on port 22 is discarded simply because the cost is higher than the direct path. Port 21 becomes the root port.

Access-2 analysis

This device will receive BPDUs on ports 21 and 22. Access-2 must decide which port is the best one. A similar decision will be made as with Access-1: port 21, the direct path, offers the best path, so it will become the root port.

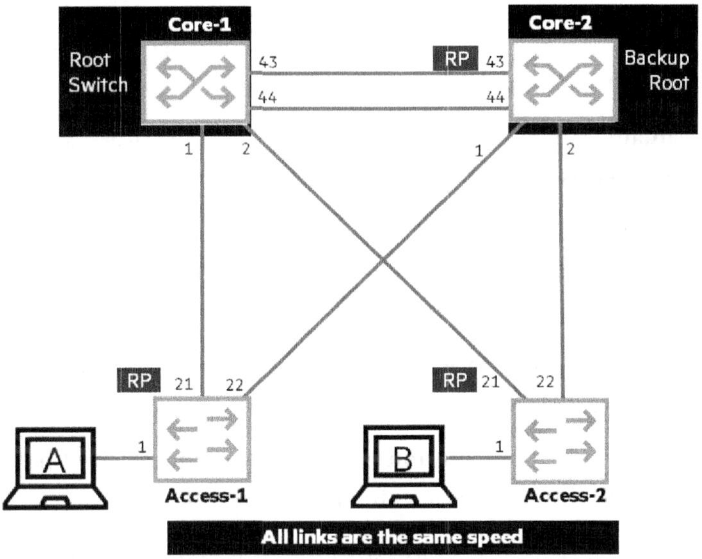

Figure 4-19: Select root ports in non-root switches

Select designated ports and alternate ports

The criteria to select a designated port is the same as the root port.

- Lowest root bridge ID
- Lowest path cost to the root bridge
- Lowest sender bridge ID
- Lowest port priority
- Lowest port ID

The following statements are a shortcut to determine a designated port on a link quickly.

- All ports in the root switch are always designated ports.
- In a switch-to-switch link where previously a root port was elected on one side, the other side must always be a designated port.

Core-1 analysis

This is the root switch, and its ports will always be the closest to the root bridge. Therefore, ports 1, 2, 43, and 44 on Core-1 become designated ports.

Core-2 analysis

The analysis focuses on ports 1, 2, and 44.

Port 1: Core-2 will evaluate if this port is closest to the root. This process is done by comparing the BPDU received from Access-1 on this port and the one that it sends out. This process follows the same rules as the root port, so let's start the analysis.

- Lowest root bridge ID. All devices in the topology agree that Core-1 is the root; this criterion does not break the tie.
- Lowest path cost to the root bridge. Access-1 and Core-2 advertise the same cost to reach the root switch (both are one hop away).
- Lowest sender bridge ID. This criterion breaks the tie since Core-2 has a lower BID than Access-1. Port 1, in this link, becomes the designated port.

Port 2: Core-2 will evaluate if this port is closest to the root. This process is done by comparing the BPDU received from Access-2 on this port and the one that it sends out.

- Lowest root bridge ID. All devices in the topology agree that Core-1 is the root; this criterion does not break the tie.
- Lowest path cost to the root bridge. Access-2 and Core-2 advertise the same cost to reach the root switch (both are one hop away).
- Lowest sender bridge ID. This criteria breaks the tie since Core-2 has a lower BID than Access-2. Port 2, in this link, becomes the designated port.

Port 44: This port cannot be designated since the port on the other side of the link is closest to the root.

Access-1 analysis

This analysis focuses on ports 1 and 22.

Port 1: This port is the only RSTP speaker in the link; this port becomes the designated port.

Port 22: This port cannot be designated since the port on the other side of the link is closest to the root.

Access-2 analysis: This analysis focuses on ports 1 and 22.

Port 1: This port is the only RSTP speaker in the link; this port becomes the designated port.

Port 22: This port cannot be designated since the port on the other side of the link is closest to the root.

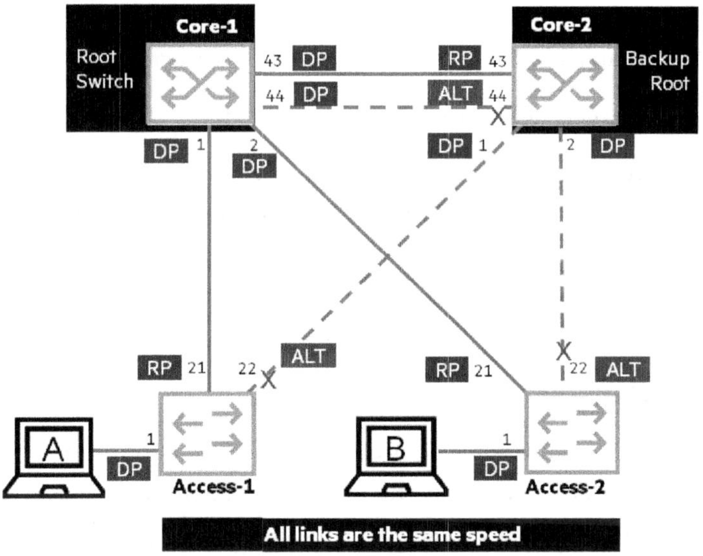

Figure 4-20: Topology with alternate ports only

The topology in **Figure 4-20** does not include any Layer 1 hubs, which implies that no backup port exists, only alternate ports. Simply put, all ports that were not elected as designated or root ports will become alternates, and their state will be Discarding. Notice that Access-1's port 22, Access-2's port 22, and Core-2's port 44 become alternate ports.

Edge ports and link types

In RSTP, an edge port is a switch port configured to immediately transition to the forwarding state without waiting for the spanning tree convergence process. These ports are typically connected to end-user devices, where there is no risk of creating network loops. Edge ports are essential for reducing network convergence time and ensuring quick connectivity for end devices.

Edge ports

Edge ports connect to endpoints, and because of this, they should not receive BPDUs. They do not need to participate in the spanning tree algorithm because they cannot cause a loop. Essentially, they are "leaves on the spanning tree." Since they cannot cause loops, they can quickly transition to a forwarding state, with no intermediate steps.

If BPDUs are received on an edge port, then the port will act as a normal spanning tree port and participate in the algorithm to prevent Layer 2 loops.

AOS-CX admin-network (default). With this the port looks for BPDUs for the first 3 seconds after the link is up. If no BPDUs are received, the port becomes an edge port and immediately forwards frames. If BPDUs are detected, the port becomes as a non-edge port. It participates in normal STP operation.

```
Switch(config)# interface 1/1/1
Switch(config-if)# spanning-tree port-type admin-network
```

An alternative to the admin-network option is the admin-edge

This option is useful in situations where you have complete control over the network topology and want to optimize the performance of specific ports connected to end devices.

You can manually configure edge ports, as shown here:

```
Switch(config)# interface 1/1/1
Switch(config-if)# spanning-tree port-type admin-edge
```

Topology change mechanism

In RSTP, a topology change occurs when non-edge ports move to a different state, or when there is a change in the state of a port that affects the network's topology. Upon detecting a change, RSTP initiates a rapid convergence process, disseminating Topology Change Notifications (TCNs) to neighboring switches to update the spanning tree topology swiftly and minimize network downtime.

In RSTP, a topology change occurs when non-edge ports move to a different state, or when BPDUs are no longer received. A switch detects this topology change, then actively informs the rest of the switches in the network of the topology change. It sets the Topology Change (TC) bit in BPDUs and transmits these BPDUs, then flushes the MAC address table entries associated with all non-edge ports.

Other switches receive a BPDU with the TC bit set from a neighbor. The switch clears the MAC address table entries learned on all its ports, except the ingress port for the TC BPDU. These switches, in turn, send BPDUs with the TC set.

In the original 802.1D standard, any switch could send TC BPDUs to notify others, but the instruction to clear the MAC address table must always come from the root switch. This two-step process takes more time to complete.

Consider the following to understand how a topology change is handled by RSTP:

1. A link failure occurs between Core-1 and Access-1.
2. As soon as Access-1 detects that its root port is down, it moves its alternate port 22 to be the new root port. Port 22 immediately moves to a forwarding state. The rest of the ports on Access-1 are unaffected.
3. Access-1 generates a TC BPDU and sends it out port 22.

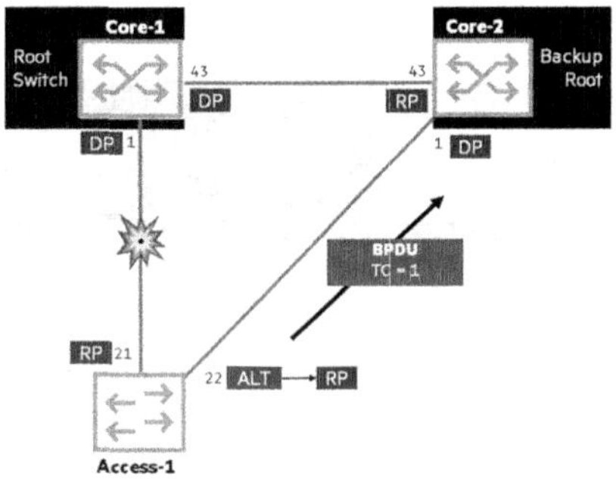

4. Access-2 flushes its MAC address tables.
5. Access-2 sends a TC BPDU.
6. Core-1 receives the TC BPDU and flushes its MAC address table.

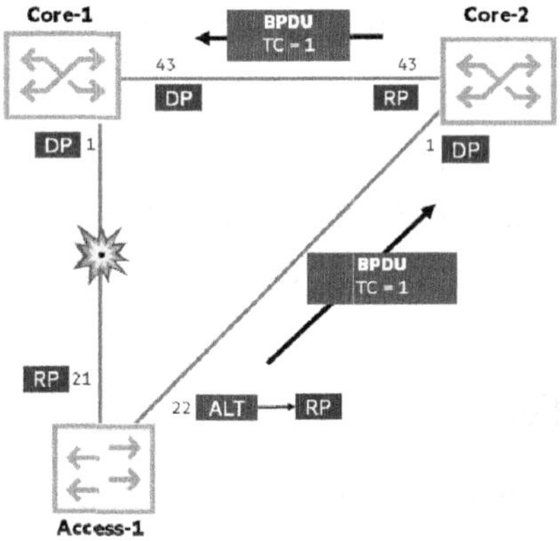

Core-1 also detects the link failure and does the same process as Access-1, generating a TC and advertising it to its neighbors.

Multiple Spanning Tree Protocol

Multiple Spanning Tree Protocol (MSTP) is a standardized protocol that enables the creation of multiple spanning tree instances within a single network infrastructure. By dividing the network into multiple spanning trees, MSTP allows for more efficient resource utilization and improved network performance compared to traditional Spanning Tree Protocols. It offers load balancing while maintaining a low CPU usage. Optimization is based on the number of possible loop-free topologies. The solution to the previous problem (single topology tree) can be accomplished by protocol IEEE 802.1s. MSTP builds an additional, new topology or instance of the Spanning Tree Protocol. Instances then can carry different VLAN traffic allowing load balancing and maintaining optimal paths.

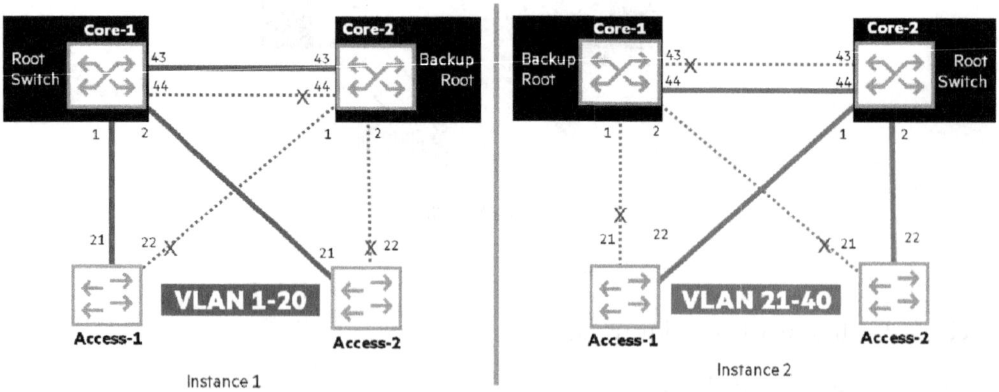

Figure 4-21: MSTP protocol reflecting two topologies on the same physical network

On the left side of the figure, VLANs 1-20 all share a single RSTP iteration. Core-1 is the root, and the tree grows from the root. VLANs share these forwarding links, with the unused links available for redundancy.

On the right side of the figure, VLANs 21-40 share a different RSTP iteration. Core-2 is the root, and the topology emanates from that root. The solid links are shared by VLANs 21-40, and the other links are available for backup. The dormant links for VLANs 1-20 are the active links for VLANS 21-40, and vice versa.

MSTP regional configuration

In AOS-CX, use the following steps to configure MSTP:

1. Configure the spanning tree mode as MSTP. The `spanning-tree mode` command sets the mode to either MSTP (open standard protocol) or Rapid Per VLAN Spanning Tree (RPVST) (Cisco proprietary).

```
Switch(config)# spanning-tree mode mstp
```

> To change the spanning tree mode, you must first disable the spanning tree with the `no spanning-tree` command in the global context.

2. Set up the MST region name. The spanning-tree config-name command sets the configuration name for the MST region in which the switch resides. All switches within an MST region must have identical configuration names. Note that the name that you provide is case sensitive.

```
Switch(config)# spanning-tree config-name <Config-name>
```

3. Set up the configuration revision number. Use the revision number to differentiate between region configurations. This is valuable, for example, when changing configuration settings within a region, and you want to track the configuration version you use. It is also valuable when you create a new region from a subset of switches in a current region and you want to maintain the same region name. All switches within an MST region must have identical revision numbers.

```
Switch(config)# spanning-tree config-revision <revision-number>
```

4. Set up the MSTP instance and map the VLAN IDs to it. The spanning-tree instance command creates a new instance with the VLAN mapped to an existing instance. Each instance must have at least one VLAN mapped to it. When VLANs are mapped to an instance, they are automatically unmapped from the instance they were mapped to before. Any MSTP instance can have all the VLANs configured on the switch.

```
Switch(config)# spanning-tree instance <instance-id> vlan <vlan-ids>
```

5. Optionally, set the switch priority for an instance.

```
Switch(config)# spanning-tree instance <instance-id> priority <priority-
value>
```

The same priority values can be used as in RSTP; the range to set up is 0-15.

To enable spanning tree globally:

```
Switch(config)# spanning-tree
```

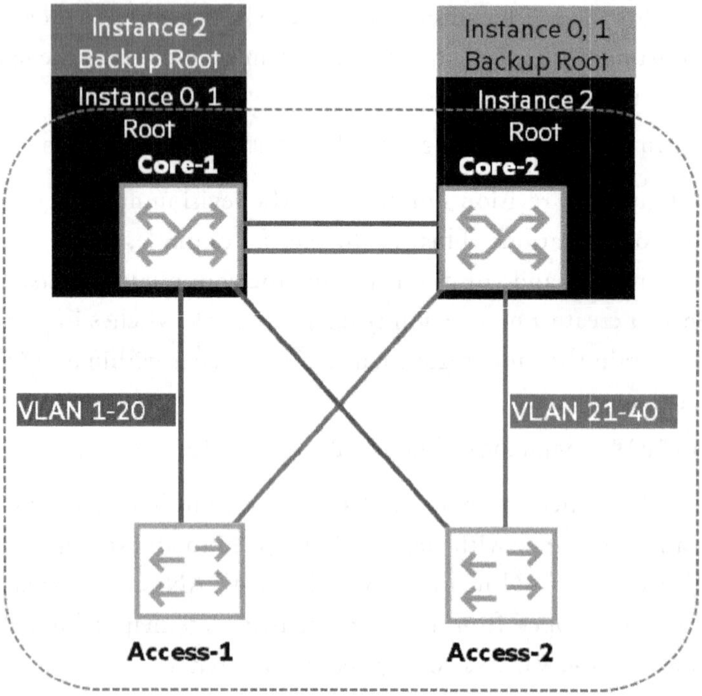

Figure 4-22: Instance 1 and 2 created for MSTP topology

CLI for common configuration for all switches

```
[config]# spanning-tree config-name CXF
[config]# spanning-tree config-revision 1
[config]# spanning-tree instance 1 vlan 1-20
[config]# spanning-tree instance 2 vlan 21-40
```

CLI for Core-1 configuration

```
[config]# spanning-tree priority 1
[config]# spanning-tree instance 1 priority 1
[config]# spanning-tree instance 2 priority 2
```

CLI for Core-2 configuration

```
[config]# spanning-tree priority 2
[config]# spanning-tree instance 1 priority 2
[config]# spanning-tree instance 2 priority 1
```

Common Spanning Tree

Common Spanning Tree (CST) is a protocol enhancement introduced in IEEE 802.1Q that addresses the limitations of spanning tree.

CST defines a single topology tree for the entire network, regardless of the number of VLANs. All BPDUs use untagged VLAN 1, and so VLANs are not considered.

The problem is that there is no load balancing, so you get sub-optimal pathing, poor link utilization, and potentially slower performance. Multiple Spanning Tree is created to resolve these issues.

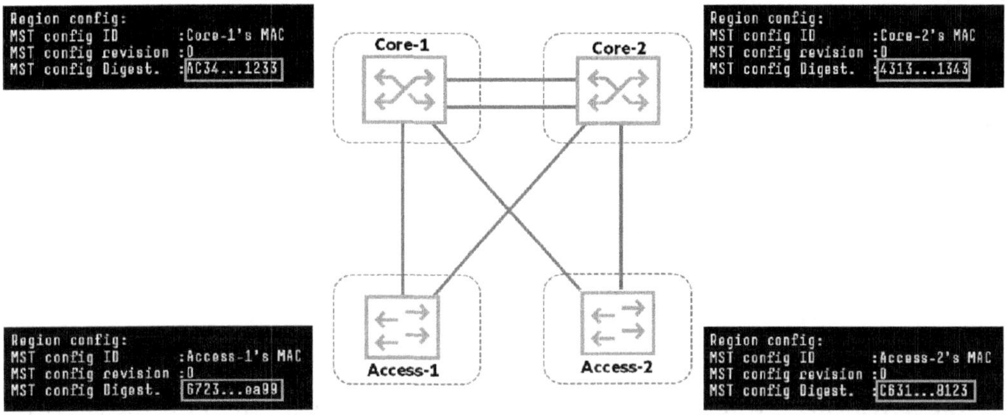

Figure 4-23: Common spanning tree configuration

In the figure, notice that each switch has MSTP enabled, but the digest or hash is different. Therefore, these devices are in different regions. For MSTP to function properly, every switch must have the same MST config ID. You can see that Core 1 and 2, Access 1 and 2 all have different Core IDs. Also, all the switches have different configuration MST config digest numbers as well. What this indicates is that instead of working as one MSTP, each switch is considering itself its own region because there is no load balancing.

By default, each switch has its own configuration ID (also known as the regional name), and the instance is 0 by default as well.

MSTP introduces the concept of a region. A region is a group of switches that have been configured with the same instance, which is mapped or associated with a group of VLANs. They must have the same MST instance name, MST revision number, and VLAN mapping. To make sure all devices have mapped the same VLANs to the instance, MSTP uses a hash algorithm (sometimes referred to as a digest).

> A hash algorithm is a mathematical function that is applied to an input (x). A hash function is interesting because from the output (y), you cannot derive or infer what the input was. Thus, it is referred to as a one-way function. Another characteristic of a hash is that if the same input (x) has been entered, the result will always be the same. MSTP takes advantage of this characteristic.

Instance 0 or Internal Spanning Tree (IST) is used as both a regular instance in MST and the creation of the CST in a multi-region deployment for backward compatibility with RSTP speakers. For this reason, Instance 0 is known as CIST (Common Internal Spanning Tree).

Learning check

Check your understanding of this material by answering the following questions.

Question 1

What is the primary purpose of Spanning Tree Protocol (STP) in Ethernet networks?

- ○ Load balacing
- ○ Loop prevention
- ○ Quality of Service (QoS) management
- ○ Network segementation

Question 2

Which of the statements is true about Rapid Spanning Tree Protocol (RSTP)? (Select two)

- ☐ It uses the same port roles and states as STP.
- ☐ It has a faster convergence time than STP.
- ☐ It supports only a single spanning tree instance.
- ☐ It operates on a per-VLAN basis.

Question 3

What is the advantage of Multiple Spanning Tree Protocol (MSTP) over traditional STP?

- ○ It supports load balancing across multiple instances.
- ○ It requires fewer configuration changes in large networks.
- ○ It has a slower convergence time than STP.
- ○ It can only used in networks with a single VLAN.

5 Link aggregation

Welcome to the link aggregation chapter, where we explore the power of combining multiple physical links into a single logical link for enhanced network performance and redundancy. In this chapter, you will discover how Link Aggregation Group (LAG) technology allows network administrators to increase bandwidth, improve fault tolerance, and optimize network efficiency by distributing traffic across aggregated links. Through practical examples and hands-on lab exercises, you will gain the skills needed to configure and manage LAGs effectively, empowering you to build robust and resilient network infrastructures.

OBJECTIVES

After completing this chapter, you should be able to:

- ✓ Identify interface requirements for LAG implementation in a network environment.
- ✓ Apply LAG configuration on network switches.
- ✓ Verify and validate load balancing effectiveness using network monitoring tools.

Overview and interface requirements

Have you ever encountered a scenario where your organization's network experience sluggish performance during peak hours due to heavy congestion? It is not hard to imagine multiple users simultaneously accessing bandwidth-intensive applications like videoconferencing, file transfers, and data backups, causing bottlenecks and slowdowns in network response times. In such situations, the need for link aggregation becomes evident, as aggregating multiple network connections into a single logical link can significantly increase bandwidth capacity, distribute traffic more efficiently, and alleviate congestion issues, ensuring smooth and uninterrupted network operations for all users.

Link aggregation, also known as link bundling or port trunking, is a networking technique used to combine multiple physical network connections into a single logical link.

In link aggregation, multiple network interfaces on a device, such as a switch or a server, are grouped together to form a single virtual interface. This virtual interface behaves as if it were a single, higher-bandwidth link, allowing for the simultaneous transmission of data across all aggregated links.

Benefits of LAG

There are three main benefits to link aggregation groups:

- **Increased bandwidth:** Aggregating multiple links effectively increases the overall bandwidth available for data transmission.
- **Load sharing:** Traffic can be distributed across the aggregated links, preventing any single link from being a bottleneck and ensuring optimal utilization of available bandwidth.
- **Redundancy and fault tolerance:** If one link in the aggregation fails, traffic is automatically rerouted through the remaining links, providing resilience and minimizing downtime.

Link aggregation comparison

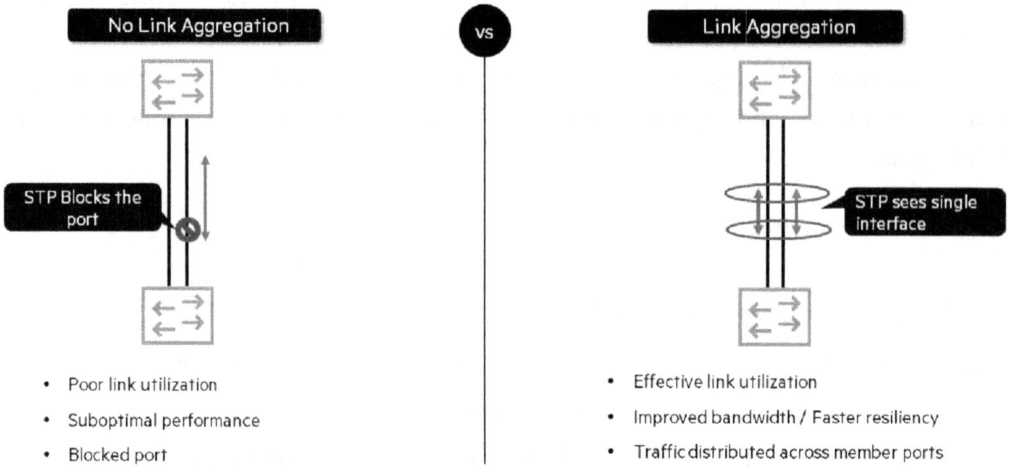

Figure 5-1: Topology comparison of no link aggregation compared to link aggregation

Switches provide connectivity for many devices on many different VLANs. All that traffic must share the switch-to-switch links. To improve performance, multiple switch interconnects must be added. While this does add some resiliency, it does not add additional bandwidth—the two links would create a loop, and so Spanning Tree Protocol (STP) will automatically block one of the ports. There are two links, but only one of them is used, leading to poor link utilization, no load balancing, and suboptimal performance.

The solution is link aggregation or multichassis link aggregation.

Link aggregation bundles multiple physical interfaces into a single logical interface. Since protocols like STP perceive this bundle as a single interface, there is no blocking. All switch interconnects carry traffic. It may go from one 10Gbps link, to two, four, or more bundled links. There is more bandwidth because traffic is distributed across member ports. There is also better resiliency because if one member fails, the remaining links carry the load. Convergence is faster than spanning tree because there is no need for an STP Topology Change notification.

Multichassis link aggregation extends the concept of link aggregation across multiple network devices, typically within the same network domain or data center. It allows for the creation of a high-bandwidth and redundant connection between devices across multiple chassis or switches.

> Link aggregation can be used not only for switch-to-switch links, but it can be used on other links to servers and routers.

Link aggregation structure

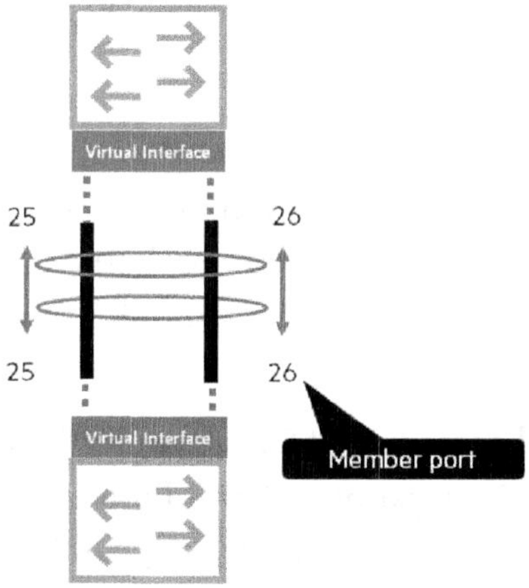

Figure 5-2: Link aggregation structure topology

When link aggregation is enabled on a switch, the protocol creates a virtual interface, then configures physical ports to be members of that virtual interface. The switch's various protocols and processes will only perceive and refer to the virtual interface. They no longer perceive the individual physical interface members.

In AOS-CX, the virtual interface is referred to as a LAG (Link Aggregation Group) and the interfaces are called member ports.

It is important to mention that broadcast and multicast traffic is sent across only one physical link in the bundle. This behavior ensures that link aggregation does not create a Layer 2 loop.

Interface requirements

The interface requirements for configuring a LAG vary depending on the specific network equipment and technologies being used. However, some common interface requirements for LAG implementation typically include:

- **Supported switch models:** Ensure that the network switches or devices being used support LAG functionality. All CX switches support LAGs. If you are using a combination of

CX switches and other brands, ensure the other brands support LAGs. If they do not, it cannot aggregate.

- **Number of ports:** Determine the number of physical ports required for creating the LAG. Most LAG configurations involve aggregating two or more physical ports required for creating the LAG. Most LAG configurations involve aggregating two or more physical ports together to form a single logical link.
- **Link types:** Verify that the physical ports being aggregated are of the same type and speed. For example, if using Ethernet ports, ensure if one port is 10 GbE, the other is also 10 GbE.

Interface best practice

- **Port configuration:** Configure the physical ports that will be part of the LAG with consistent settings, such as speed, Duplex mode, and other relevant parameters. Ensuring uniform port configurations helps prevent compatibility issues and ensure optimal performance within the LAG.
- **Link Aggregation Control Protocol (LACP) support:** Determine whether the network equipment supports Link Aggregation Control Protocol (LACP), which is a standard protocol used for negotiating and managing LAGs dynamically. LACP helps ensure interoperability and seamless operation of LAGs across multiple devices.
- **Topology considerations:** Assess the network topology and ensure that the LAG configuration aligns with the overall network design and requirements. Consider factors such as redundancy, traffic distribution, and scalability when planning the LAG implementation.

By meeting these interface requirements, network administrators can effectively configure and deploy LAGs to optimize network performance, increase bandwidth, and enhance fault tolerance within their network environments.

Bottom line: If you use LACP on one side, you need LACP on the other side. If one side has specific settings for speed, duplex, and so forth, the other must be configured the same. You must MATCH the settings, speed, and so forth on each side in order to get the LAG to work properly.

Static and dynamic LAGs

Static and dynamic LAGs are two distinct approaches to aggregating multiple physical links into a single logical link within a network infrastructure. While a static LAG has manually grouped ports, a dynamic LAG uses protocols such as LACP.

Static LAG

In a static LAG configuration, as a network administrator, you must manually group together specific ports on network switching to form a LAG without relying on dynamic protocol. This manual configuration involves specifying which ports on the network switches will be grouped together to form the LAG.

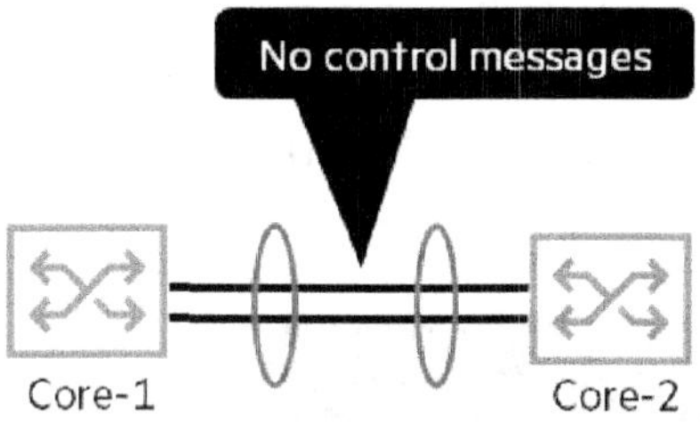

Figure 5-3: Static LAGs are connected to their peer with no control messages

Once configured, the static LAG remains fixed until manually modified by the administrator. In static link aggregation mode, devices do not exchange any control information. There is no signaling between switch peers about the LAG; you simply configure the LAGs independently of each peer. If your configuration is good, the LAG works. However, the switch peers have no knowledge of who they are connected to, or whether they are connected to the same peer. This mode is not recommended because a misconfiguration on one side is not detected by the peer. This can lead to unexpected behavior, which can be challenging to detect and troubleshoot.

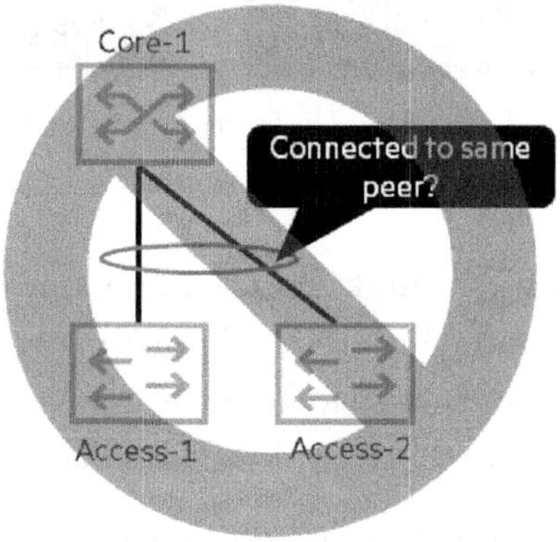

Figure 5-4: Static LAG where Access-1 is not aware of connection to Access-2

Another scenario that could cause problems is shown in the above figure example. Switch Access-1 is not aware that you have erroneously connected LAG member ports to two separate physical switches.

Static LAGs are often preferred in scenarios where simplicity and predictability are valued over dynamic negotiation. They provide a straightforward method for aggregating links without replying on additional control protocols, making them easier to deploy and manage in smaller or less dynamic network environments. However, static LAGs may lack the automatic failover and load-balancing capabilities offered by dynamic LAGs, requiring careful planning and configuration to ensure optimal performance and redundancy.

Configuring a static LAG

In a static LAG configuration, as a network administrator, you must manually group together specific ports on network switching to form a LAG without relying on dynamic protocol. This manual configuration involves specifying which ports on the network switches will be grouped together to form the LAG.

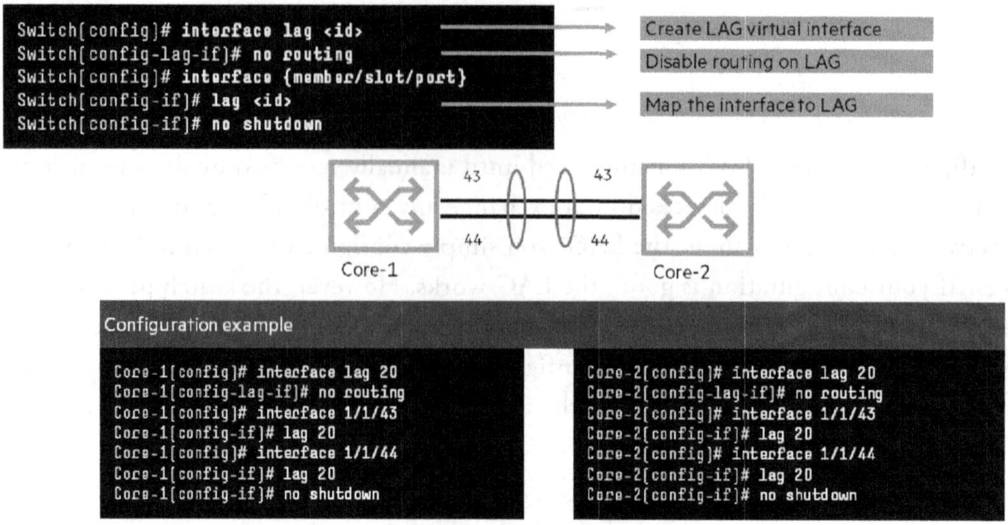

Figure 5-5: CLI code to configure static LAG

Dynamic LAG - LACP

IEEE 802.3ad, commonly referred to as Link Aggregation Control Protocol (LACP), is a standard protocol used for dynamic LAG configuration in Ethernet networks. With LACP, multiple physical links between network devices such as switches or servers can be dynamically aggregated into a single logical link, providing increased bandwidth, redundancy, and load balancing.

LACP operates by negotiating the formation of a LAG between the participating devices. When enabled on the compatible network interfaces, LACP-capable devices exchange LACP frames to automatically detect and establish link aggregation. During the negotiation process, LACP ensures that all links in the LAG have compatible settings, such as speed, Duplex mode, and VLAN membership.

Once the LAG is formed, LACP continues to monitor the status and health of the aggregated links. If a link within the LAG fails or becomes inactive, LACP can dynamically redistribute traffic across the remaining active links to maintain network connectivity and optimize bandwidth utilization. Additionally, LACP supports load balancing mechanisms to distribute traffic evenly across all aggregated links, preventing any single link from becoming overwhelmed with traffic.

Dynamic LAG - 802.3ad

As a reminder, 802.3ad contains all of the following:

Link Aggregation Control Protocol (LACP)

- Peers exchange periodic LACP data unit messages.
- Switches know about LACP peers.
- Recommended implementation.

LACP exchanges periodic messages called LACP data units

These messages include:

- System ID: Identifies the LACP peer.
- Operational key: Identifies the LAG instance.
- Include Flags: There are some flags that are included in the LACP data unit.

The LACP data unit flags

- LACP_Activity
- LACP_Timeout
- Aggregate
- Synchronization
- Collecting
- Distributing

Dynamic LAG or LACP is the recommended method for implementing link aggregation, to avoid unexpected network problems.

The amount of switch interfaces that can utilize LACP is dependent on the specific switch.

802.3ad operational modes and LACP rate

LACP can be configured in one of two modes: Passive or Active. These modes control how peer negotiation proceeds.

With **Passive mode**, the device passively waits to receive an LACP data unit message from the peer, to dynamically create the LAG. This mode places the LAG in a listening state; it is as if the switch is thinking, "I will just sit here and wait until I hear from my peer." If you configure the peer switch to also be in Passive mode, then it is thinking the same thing. Both switches wait to hear from the other, and so a functional LAG connection is never formed. At least one peer must be in Active mode.

With **Active mode**, the device actively transmits LACP data unit messages over its member ports: "Hey! Let's form a LAG." Whether the peer is Passive or Active, it responds, negotiation continues, and the LAG successfully forms.

LACP rate is used to configure the timing of the heartbeat between LACP peers. By default, AOS-CX switches operate at the **LACP rate slow**, meaning that the switch will send out a heartbeat request every 30 seconds. The timeout period is three consecutive heartbeat loss that is 90 seconds. To optimize the detection of a change or a broken link through the LACP timeout, you may configure the **LACP rate fast**, meaning that the switch will send out a heartbeat request every second and the timeout period is a three-consecutive heartbeat loss that is 3 seconds.

Device A port mode	Device B port mode	LAG formed?
Passive	**Passive**	**No**
Passive	Active	
Active	Passive	Yes
Active	Active	

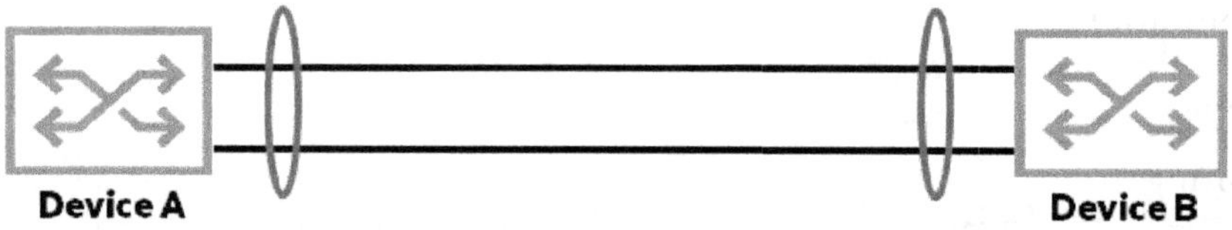

LACP rate	Heartbeat	Timeout
Slow	30 seconds	90 seconds
Fast	1 second	3 seconds

Figure 5-6: Passive and Active mode negotiations with LACP rate

Configuring dynamic LAG

LACP can be configured in one of two modes: Passive or Active. These modes control how peer negotiation proceeds.

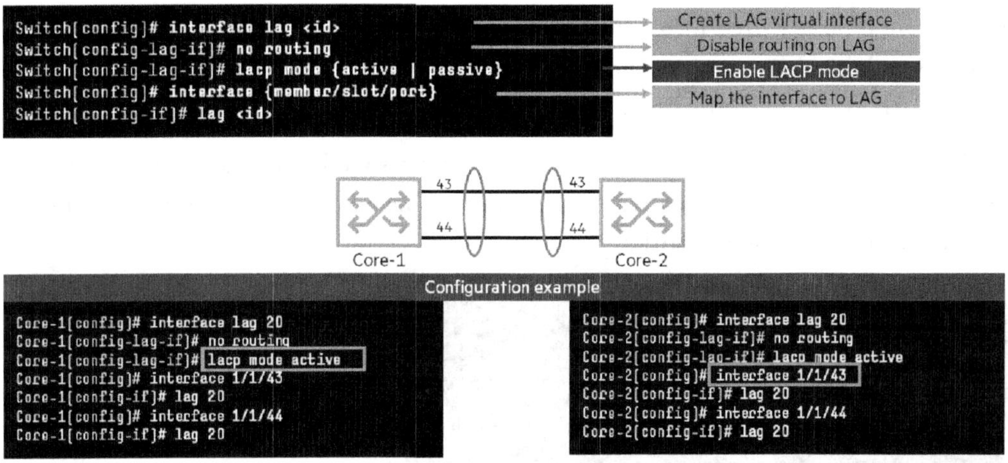

Figure 5-7: CLI code for dynamic LAG configuration

Verifying LAG interface

Use `show interface lag` to verify the dynamic or static LAG has been set up correctly. There is also a way to see the LAGs total available bandwidth utilizing the `show interface lag brief` command.

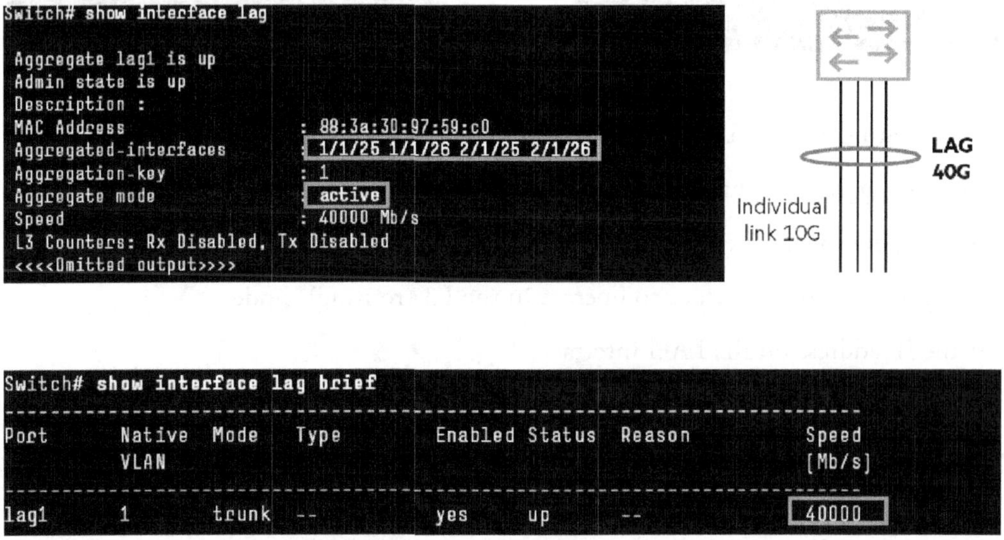

Figure 5-8: CLI code to verify LAG configuration

Static Layer 3 LAG configuration

Usually, LAGs are used for L2 traffic, but we may also use it with routed traffic (as a routed interface) if needed.

Layer 3 LAGs are typically used in network configurations where routing between different network segments or VLANs is required. Layer 3 LAGs provide a scalable and resilient solution for distributing traffic across multiple links at the network layer.

Layer 3 LAGS can be deployed between routers or Layer 3 switches to increase the aggregate bandwidth between different network segments or subnets. This enables faster routing of traffic between different parts of the network.

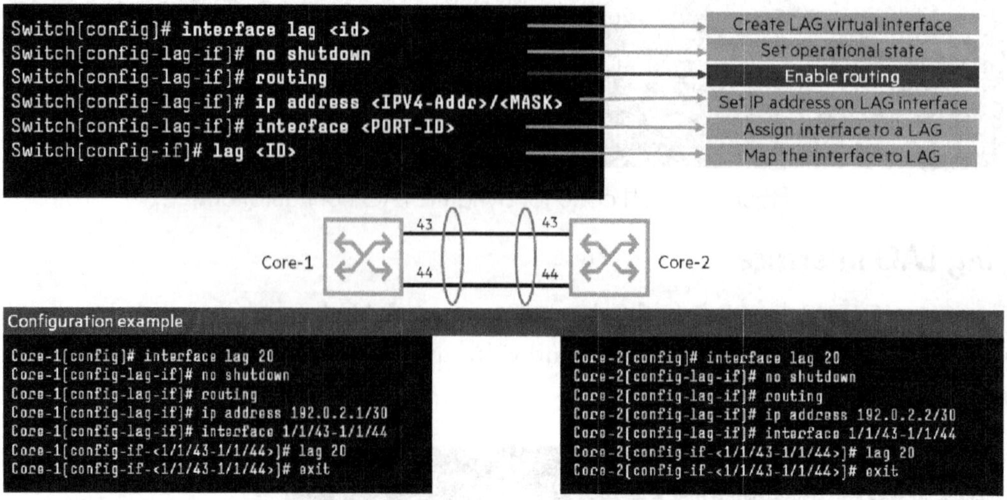

Figure 5-9: Static Layer 3 LAG configuration steps.

The steps to creating a Static Layer 3 LAG are:

1. Create an aggregated interface and access the LAG interface.
2. Set the operational state of the LAG interface.
3. Configure the LAG interface to operate in the L3 (routing) mode.
4. Set the IP address on the LAG interface.
5. Access an interface or range of interfaces to the LAG.
6. Assign (map) the LAG IF to the interface or interface range and set the administrative state of the member ports to up.
7. View your configuration.

Load sharing

Load sharing with load balancing algorithms in LAGs is a critical aspect of optimizing network performance and resource utilization. By distributing network traffic across multiple aggregated links, load balancing algorithms ensure efficient use of the available bandwidth and prevent individual links from becoming overwhelmed with traffic. These algorithms intelligently analyze factors such as link utilization, packet characteristics, and network topology to determine the optimal distribution of traffic. Whether using simple hash-based algorithms or more advanced techniques, like weighed load balancing, load sharing in LAGs plays a pivotal role in maximizing network throughput and minimizing congestion.

Load sharing algorithm

Peer LAG devices must balance the traffic load across multiple physical interfaces. How does the switch decide which member port should be used for a given packet? It uses a hash algorithm.

A hash algorithm is a mathematical function that is applied to an input (x). Interestingly, if you see the output (y) of this function, you cannot derive or infer input (x). Therefore, it is referred to as a one-way function. Another characteristic of a hash is that if the same input (x) has been entered, the result will always be the same. So, what does the switch use as inputs to this function?

The switch uses packet header information as hash function inputs, and the result is the port member to be used for that particular packet. Inputs for the hash could be:

- Layer 3 source and destination IP addresses
- Layer 2 source and destination MAC addresses

The figure shows how the Layer 2 source and destination MAC addresses are taken as inputs for the algorithm. The output is to use physical port 1 for all packets with this source and destination MAC address combination—always. Only a link failure will cause port 2 to be used for these packets.

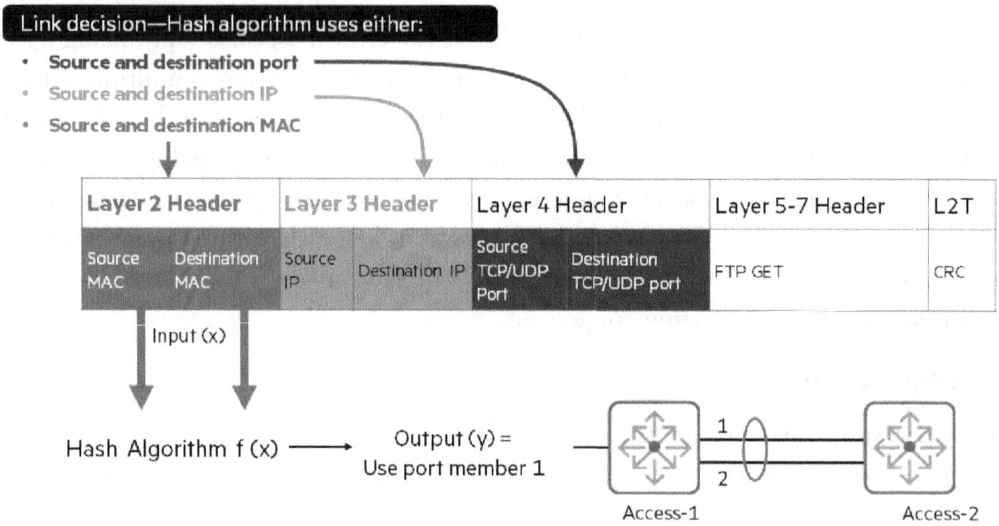

Figure 5-10: Algorithm from Layer 2 and/or Layer 3 headers

Verify load sharing

In AOS-CX, the default input information for the hash algorithm is the source and destination IP addresses. You can verify with the `show lacp aggregates` command. The highlighted hash information is `l3-src-dst`. L3 means Layer 3, which is the IP addresses.

In some cases, the hash algorithm's use of source/destination IP addresses may not properly distribute the load equally among port members. This could lead to traffic congestion and poor performance.

In that case, you can modify the hash input data. As shown in the figure, you have decided to use Layer 2 source and destination MAC addresses as inputs to the hash function. This might help to avoid port member oversubscription.

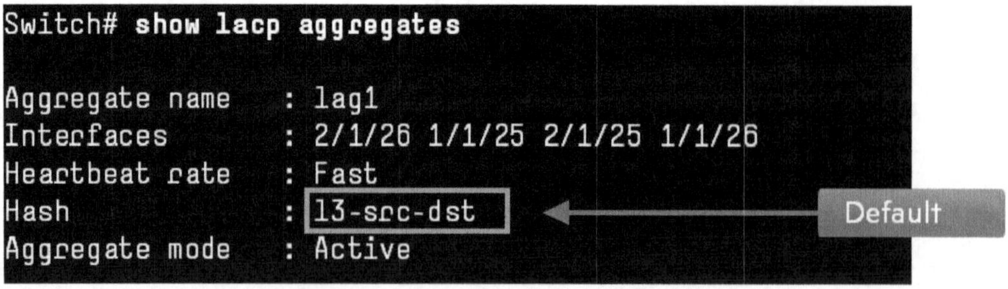

Figure 5-11: CLI to verify load balancing

```
Switch(config)# interface lag <ID>
Switch(config-lag-if)# hash l2-src-dst
```

Figure 5-12: CLI to change hash algorithm to L2

LACP data unit flags

LACP data units use various flags to convey information during the negotiant and management of link aggregation between network devices. These flags include:

- **Activity (ACT):** The ACT flag indicates whether a network device is actively sending LACP packets. When the ACT flag is set, it signifies that the device is actively participating in LACP negotiations. If the flag is cleared, it indicates that the device is in a passive state, only receiving LACP packets.
- **Timeout (TIME):** The TIME flag specifies the timeout period for LACP negotiations. It can be set to short or long timeouts, depending on the desired responsiveness of the LACP protocol. Short timeouts lead to faster detection of link failures, while long timeouts reduce the frequency of LACP packet exchanges.
- **Aggregation (AGG):** The AGG flag indicates whether a link aggregation group (LAG) has been formed between the network devices. When the AGG flag is set, it signifies that the devices have successfully negotiated the formation of a LAG and are actively aggregating their links. If the flag is cleared, it indicates that the devices are not yet part of a LAG.
- **Synchronization (SYNC):** The SYNC flag is used to synchronize LACP information between the participating devices. When the SYNC flag is set, it signifies that the devices have exchanged LACP information and are in sync with each other's configuration. If the flag is cleared, it indicates a lack of synchronization, prompting devices to exchange LACP information.
- **Collecting (COL):** This flag indicates that a network device is actively collecting information about the LAG from its partner device. When the COL flag is set, the devices is receiving LACP data units containing information about the aggregated links, including their status, speed and other parameters. This information I crucial for the device to synchronize its view of the LAG with its partner.
- **Distributing (DIST):** The Distributing flag indicates that a network device is actively distributing information about the LAG to its partner device. When the DIST flag is set, the device sends LACP data unites containing information about its own aggregated links, including their status and capabilities. This allows the partner device to synchronize its view of the LAG and make informed decisions about link aggregation and load balancing.

By utilizing these flags, LACP enables network devices to exchange information about link aggregation status, negotiate LAG information, and ensure synchronization of LACP parameters, facilitating the dynamic management and control of link aggregation in Ethernet networks.

Table 5-1: LACP data unit flags

Flag	Meaning
LACP_Activity	Indicates a participant's intent to detect or maintain the LAG. ■ Set = Active LACP ■ Not Set = Passive LACP
LACP_Timeout	The participant wishes to receive periodic transmissions. ■ Set = Short timeout ■ Not Set = Long timeout
Aggregate	LAG has been formed between the network devices
Synchronization	Devices have exchanged LACP information and are in sync with each others configuration.
Collecting	Local switch is actively collecting information about the LAG from its partner device
Distributing	Local switch is actively distributing information about the LAG to its partner device

Learning check

Check your understanding of this material by answering the following questions.

Question 1

Which IEEE standard defines the Link Aggregation Control Protocol (LACP)?

- O IEEE 802.1Q
- O IEEE 802.3ad
- O IEEE 802.11
- O IEEE 802.15.4

Question 2

What is a characteristic of Static Link Aggregation Group?

- O It utilizes protocols like LACP for automatic link aggregation.

- The aggregation of links is manually configured by network administrators.
- It dynamically negotiates and manages the formation of LAGs.
- Static LAGs provide automatic failover and load-balancing capabilties

Question 3

What is the primary purpose of load balancing algorithms in LAGs?

- To increase the number of physical links in the LAG.
- To prioritize certain types of traffic over others.
- To distribute network traffic evenly across aggregated links.
- To establish redundant connections for fault tolerance.

6 Switch stacking and

In this chapter, we will explore the concept of switch stacking, a technique used to interconnect multiple switches to function as a single, unified entity. There are benefits, configuration, and best practices associated with switch stacking, which empower you to optimize your network infrastructure for scalability, redundancy, and manageability.

OBJECTIVES

After completing this chapter, you should be able to:

- ✓ Differentiate between operational planes.
- ✓ Evaluate stacking and virtualization technologies.
- ✓ Demonstrate the application of Virtual Switching Framework (VSF) technology.

Introduction

Switch stacking is an important concept to understand. Stacking in networking refers to the process of interconnecting multiple network switches to operate as a single, unified entity. This allows switches to function together as if they were a single switch, providing increased scalability, simplified management, enhanced redundancy, and improved performance.

Stacking technologies

What are stacking technologies?

Stacking technologies refer to the methods used to interconnect multiple network switches. These technologies facilitate the creation of a stack, allowing switches to function together seamlessly and providing various benefits.

For stackable switches, the stack is usually built with cables that connect all the switches in a specific topology.

Those cables are connected to specific ports of the switches. Frontplane stacking uses standard Ethernet ports on the front of the switches and can utilize standard cables to connect these ports to build the stack. Backplane stacking involved interconnecting switches through an internal backplane connector, typically located at the rear of the switches. This method often uses proprietary connectors and cables. VSF is a specific implementation of frontplane stacking, because it uses the front interface to stack the switches.

Stacking benefits

Switch stacking offers several benefits:

Scalability: Combining multiple switches into a single logical unit increases ports and bandwidth, simplifying network expansion without complex configurations.

Simplified Management: Administrators can manage multiple switches as one entity, reducing administrative tasks and enhancing efficiency.

Enhanced Redundancy: Stacked switches offer built-in redundancy with features like link aggregation, ensuring high availability in case of failure. By combining multiple switches into a single logical unit, STP (Spanning Tree Protocol) issues can be eliminated, further enhancing network reliability and reducing the risk of downtime.

Improved Performance: Load balancing and traffic distribution across switches optimize network performance and reduce bottlenecks and latency.

Potential Cost Savings: Eliminating redundant infrastructure and simplifying cabling requirements results in savings on equipment, deployment, and maintenance costs.

Port numbering

Member: When HPE Aruba Networking Virtual Switching Framework (VSF) or HPE Aruba Networking Virtual Switching eXtension (VSX) features are running, multiple switches can be seen as one virtual switch. This member number indicates the member ID in the cluster. By default, an AOS-CX switch will be member 1, indicating that these features are not running.

Slot: In modular switches like the CX 8400 Series and CX 6400 Series, this number represents the slot being used by a particular line card. For fixed switches (CX 8320 Series, CX 8325 Series, and CX 6300 Series), this number will be always 1.

Port: This number refers to the individual interface in the line card (modular switches) or in the chassis (fixed switches).

VSX versus VSF

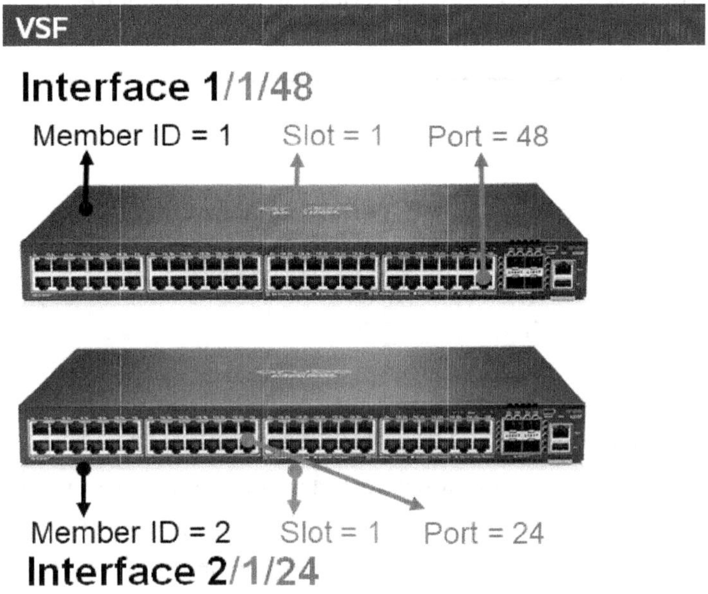

Figure 6-1: VSF port numbering examples of multiple interfaces

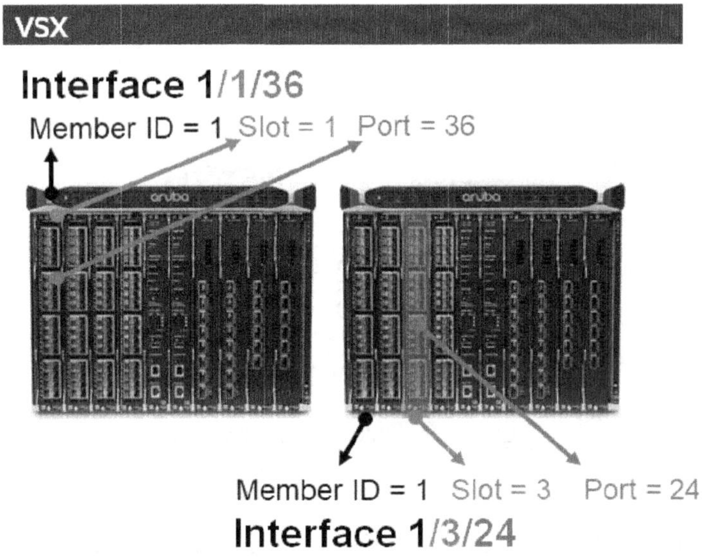

Figure 6-2: VSX port numbering examples

In a stack of switches using VSF technology, the member ID is used to identify the switch member within the stack. On the other hand, VSX is a switch virtualization technology that enables each switch's management and control planes to operate independently. Consequently, VSX switches will always have their member ID set to one, even if they are configured as part of a VSX.

VSF is available for the CX 6200 Series and CX 6300 Series. VSX is supported on the other platforms.

Operation planes

Operation planes in networking refer to distinct functional layers within a network device that handle different aspects of operation. These planes include the data plane, which forwards traffic. The control plane manages routing and switching decisions, and the management plane handles device configuration and monitoring.

Data planes

The data plane, also known as the forwarding plane or forwarding engine, is responsible for the actual forwarding of data packets within a network device. It operates at the lowest layers of the OSI model and performs tasks such as packet switching, routing, and filtering based on pre-defined rules and forwarding tables. Essentially, the data plane forwards the incoming frames based on their destination address, ensuring that data is efficiently transmitted across the network.

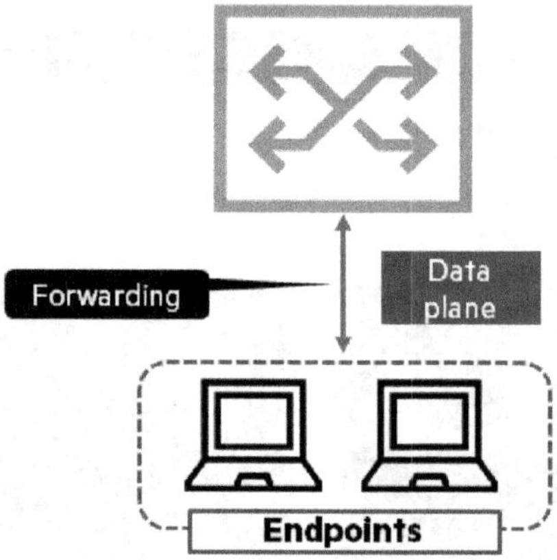

Figure 6-3: Data plane topology

Control planes

The control plane in networking is responsible for managing the overall operation and behavior of a network device. It handles tasks such as routing protocol operations, updating routing tables, and making forwarding decisions based on network topology and policies. Essentially, the control plane is responsible for building the tables that the data plane will use and how data is forwarded. It operates independently of the data plane and is crucial for ensuring proper network operation connectivity but has a tight relationship with the data plane.

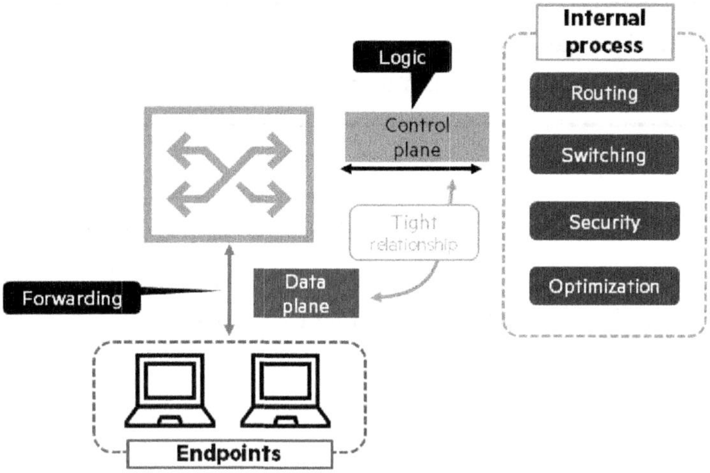

Figure 6-4: Control plane and data plane topology

Management planes

The management plane in networking is responsible for device configuration, monitoring, and maintenance. It provides a means for network administrators to interact with and control network devices, typically through a management interface such as CLI, a web-based interface, or Simple Network Management Protocol (SNMP). The management plane enables administrators to configure device settings, monitor device performance and status, troubleshoot network issues, and implement security policies. It operates independently of the data and control planes, focusing solely on administrative tasks related to device management and operation. Control and management plane functions are centralized in one group member, but each member runs its own independent data plane. The tight relationship between the control and data planes is maintained; it just happens on an inter-switch basis.

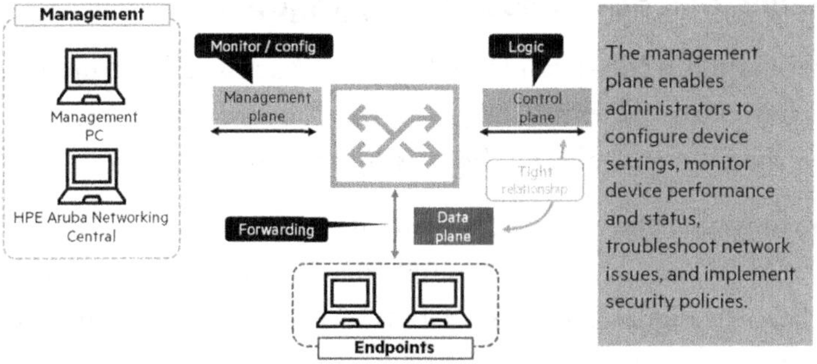

Figure 6-5: Management, control, and data plane topology

AOS-CX devices have a specific interface and VRF that are used for Out-of-Band Management, which maintains a total separation from the data plane.

Virtual switch

Stacking technology allows you to manage a group of switches as a single device—a virtual switch.

A virtual switch is a group of switches managed as a single virtual device. It features a centralized control and management plane as well as independent data planes.

Stacking benefits include:

- Ease of management:

  You no longer need to connect to, configure, and manage each individual switch. You simply configure the primary switch. That configuration is then automatically distributed to other virtual switch members. This simplifies network setup, operation, and maintenance.

- Network simplification:

  Since multiple devices share a common control plane, routing protocols and spanning tree are no longer needed inside the stacking group. Connected devices perceive the group as a single device.

HPE Aruba Networking switching families supports two stacking and virtualization technologies: HPE Aruba Networking Virtual Switching Framework (VSF) and HPE Aruba Networking Virtual Switching eXtension (VSX). This training is focused on VSF; however, you will have a brief introduction to VSX.

Distributed data plane and distributed link aggregation

Stacking technologies maintain the data plane distributed across all the members. This implies each physical switch individually creates and populates forwarding tables like ARP and MAC. These tables are then shared across all members using the control plane.

Mobility controllers, firewalls, and servers can benefit from stacking with LAG-enabled switches because they perceive a LAG connection to a single device. However, physical links terminate in different stack members. In **Figure 6-6**, if SW1 fails, the traffic can still use other LAG links.

LAG and stacking features enable the network to fully use all available links at the same time. Notice that spanning tree is not needed because the stack operates from a single control plane. HPE highly recommends this implementation.

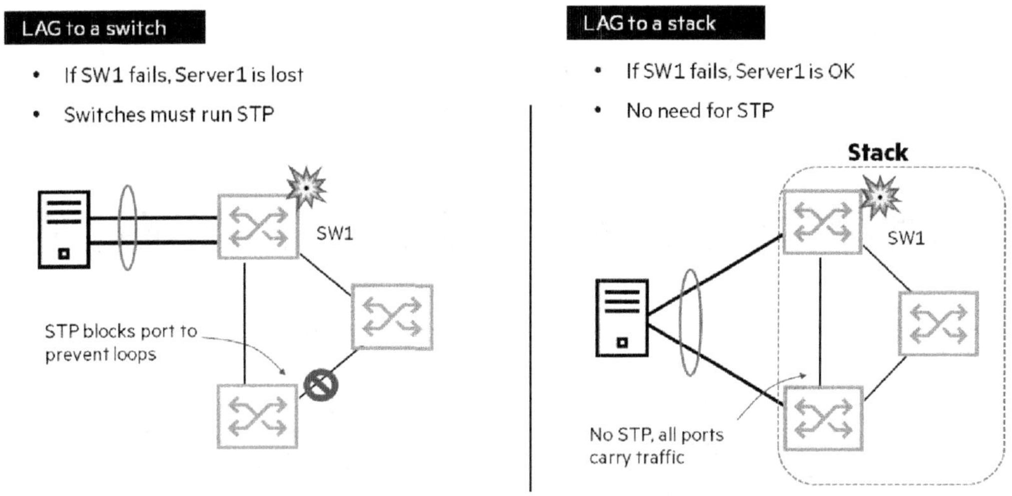

Figure 6-6: LAG to a switch versus LAG to a stack

HPE Aruba Networking Virtual Switching Framework

HPE Aruba Networking Virtual Switching Framework (VSF) is a stacking technology developed by some network equipment vendors to create a virtual switch using multiple physical switches. VSF allows network administrators to combine multiple physical switches into a single logical entity, providing high availability, scalability, and simplified management. With VSF, the aggregated switches operate as a single switch with shared control plane and management plane functionalities, enabling seamless failover and load balancing of network traffic across the stack. The technology enhances network resilience, reduces complexity, and improves performance, making it ideal for deployment in enterprise and data center environments. In VSF, each stack member has links to two other members, ensuring resiliency against link and hardware failures.

Stacking solution and platforms

VSF defines a virtual switch, which is composed of individual physical switches interconnected using Ethernet links. All member switches share a single control plane. End devices connected to VSF perceive a single device. As you recently learned, this virtual switch behavior has significant benefits related to simplified management and improved connectivity.

In AOS-CX, VSF is supported on the CX 6200 Series and CX 6300 Series models only; other models use a different technology, called VSX. The CX 6200 Series supports 8 members in a stack, and the CX 6300 Series supports a maximum of 10 members in the stack. This feature is enabled by default.

The VSF feature is also available in the HPE Aruba Networking AOS-S switching family, including the 5400 and 2930 series, but the feature is disabled by default. Understand that the VSF feature is not compatible between AOS-CX and AOS-S platforms. You cannot form a VSF stack with switches from different OS families.

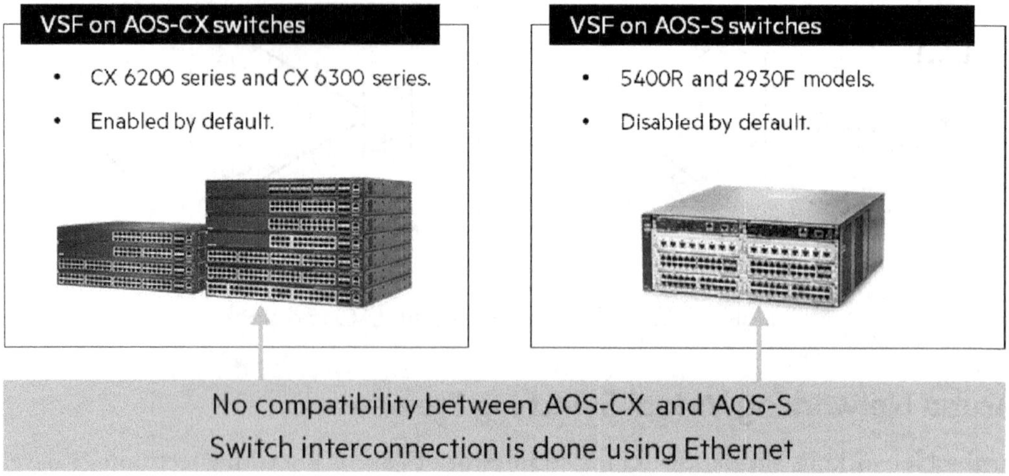

Figure 6-7: VSF on AOS-CX and AOS-S switches

Member roles and links

VSF creates a single control plane that runs on a single VSF member, sharing the information with the entire stack. This is the primary member and always uses member ID 1. This device assumes the conductor role in the stack.

A factory-default CX 6200 Series or CX 6300 Series switch boots up as a VSF-enabled switch with member ID 1. This implies that the switch behaves as the primary member.

A secondary member provides for high availability, in case of primary failure. You can choose any member in the stack to take the secondary role, but it must be explicitly defined. Configure any member ID except 1.

The other devices in the VSF are members, which only run the data plane and cannot assume the conductor role.

The CX 6300 Series VSF switches is interconnected using SFP uplink ports. When you configure a port for VSF, it can no longer be used as a Layer 2 or Layer 3 interface. In other words, the port does not belong to the switch's data plane.

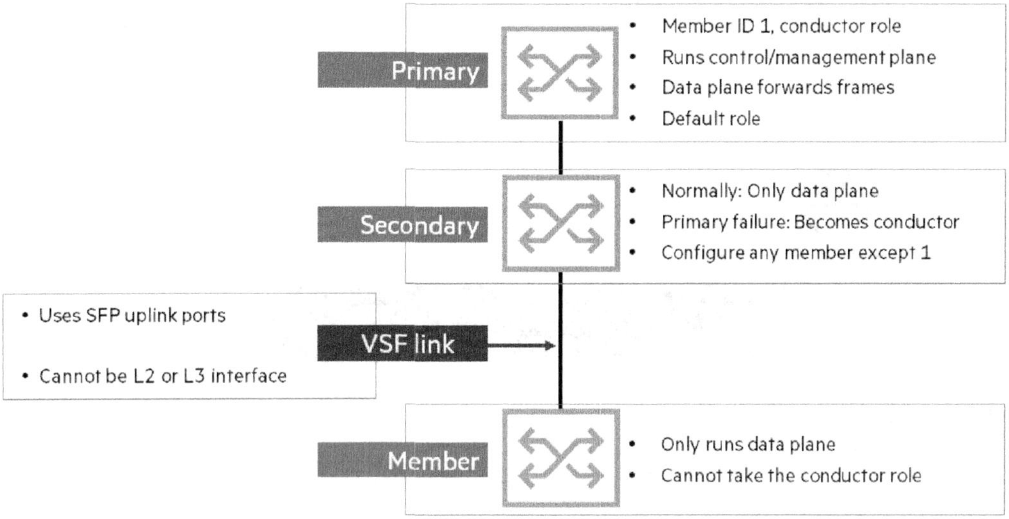

Figure 6-8: VSF member roles and links

Open virtual switch database

The objective of the VSF open virtual switch database is to track/sync state and configuration data for the stack.

The current state database is an important element of the AOS-CX architecture; it is contacted by all control and management protocols. In a VSF stack, this database run on the conductor switch.

When VSF runs on a group of switches, the Open vSwitch Database (OVSDB) is also created. This new database runs in the conductor switch and contains state and configuration data for the VSF stack itself.

The conductor switch synchronizes OVSDB content with the standby to ensure that it can quickly take over the conductor role without interruption.

The OVSDB database includes six tables:

- VSF member table
- VSF link table
- System table: Includes the number of members, Multi-Active Detection (MAD) status, fragment status, and topology type
- Subsystem table: Boot time for each member
- Interface table
- Topology table

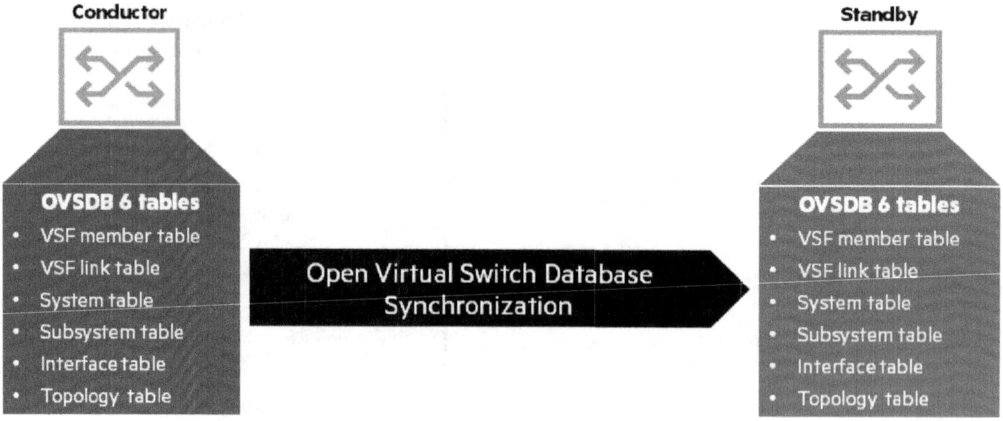

Figure 6-9: VSF open virtual switch database

VSF topologies

VSF members must be physically connected to form the switching stack in one of two topologies: Daisy chain or Ring.

- **Daisy chain topology:** As the name implies, this topology interconnects VSF members with a single chain of Ethernet connections. From the figure, you can easily see that a switch or link failure causes the stack to be split. This means that part of the stack is unable to provide endpoint connectivity.

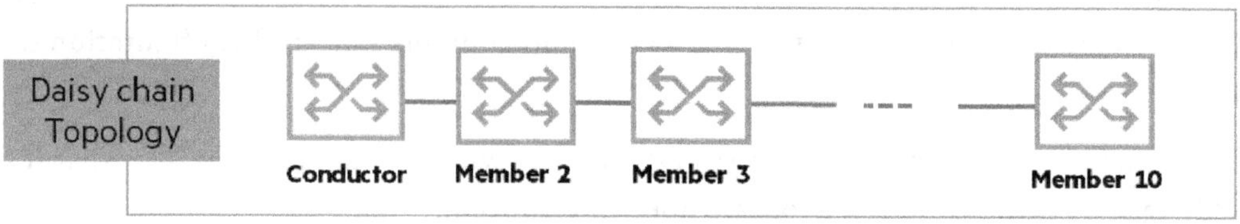

- **Ring topology:** This topology is recommended since it offers a backup path in case of a switch or link failure.

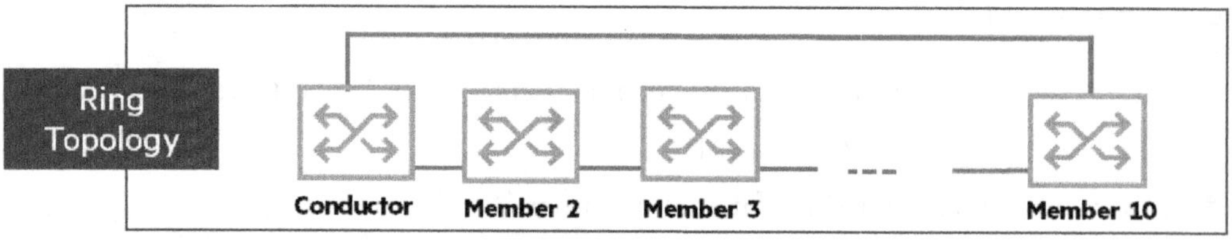

Full mesh topology is not supported.

VSF requirements

VSF has some requirements that you must meet:

- AOS-CX version 10.4 or higher.
- All members must run the same AOS-CX version.
- CX 6200 Series can stack up to 8 switches with other CX 6200 Series.
- CX 6300 Series can stack up to 10 switches with other CX 6300 Series.
- VSF logical link between a pair of switches, using a regular 10, 25, or 50 Gbps port.
- All VSF links in a stack should operate at the same speed.

Stacks cannot combine different switch families, for example, CX 6200 Series with CX 6300 Series. However, different models of the same family could be used in a stack, for example, a 24-port and a 48-port CX 6300 Series in the same stack.

CX 6200 Series

VSF allows stacks to be formed using any combination of SKUs of the CX 6200 Series family. Up to eight member switches will be allowed.

CX 6300 Series

VSF allows stacks to be formed using any combination of SKUs of the 6300 family. Up to 10 member switches will be allowed. Connections between the switches must use 10G, 25G, or 50G links. All VSF links in a stack should operate at the same speed.

VSF configuration types

In the context of VSF, the terms "manual," "pre-provisioning," and "auto-stacking" refer to different methods for configuring and managing the stack of switches.

Here is an overview of each method:

Manual stacking

Definition: Manual stacking involves manually configuring each switch in the stack. Administrators have to manually set the stack IDs, roles (such as primary or secondary), and other relevant parameters on each switch.

Use Case: This method is typically used in environments where precise control and customization of the stack configuration are necessary. It is often preferred by network administrators who need to tailor the stack settings to specific requirements.

Pros: Offers fine-grained control over the stack configuration.

Cons: Can be time-consuming and prone to human error, especially in large deployments.

Pre-provisioning

Definition: Pre-provisioning allows administrators to configure the stack settings before physically connecting the switches. This involves setting up the stack IDs, roles, and other parameters in advance.

Use Case: Useful in scenarios where the network design and switch roles are planned ahead of time, allowing for a smoother and quicker deployment process.

Pros: Reduces configuration time during deployment and minimizes the risk of configuration errors.

Cons: Requires upfront planning and may not be as flexible for dynamic or ad-hoc changes.

Auto-stacking

Definition: Auto-stacking is a more automated approach where switches automatically recognize each other and configure the stack with minimal manual intervention. When switches are physically connected, they automatically negotiate roles and stack parameters.

Use Case: Ideal for environments where ease of deployment and minimal manual configuration are priorities. It is suitable for smaller networks or scenarios where rapid deployment is necessary.

Pros: Simplifies the stack setup process and reduces the need for manual configuration.

Cons: May offer less control over specific stack configurations and might not be suitable for highly customized network setups.

Choosing the appropriate method depends on the specific requirements, scale, and complexity of the network deployment.

Manual configuration example

A couple of configuration steps are needed to configure a VSF stack of switches.

1. Select which switch you want to become the conductor (or primary) switch for the stack.
2. Then you need to configure the interface to be used as the VSF.
3. After the conductor switch is configured, it is time to configure the next switch. Start by configuring which port will be used as VSF link, and then, renumber the switch unit ID to 2. It will trigger a reboot for the switch to join the stack.

4. Once switch 2 is back online and active, add the following switch (unit 3) and so on.
5. When the stack is formed, is time to select the standby switch. Notice that it will trigger a reboot of that switch.

Consider the scenario where two CX 6300 Series are assigned to the same VSF stack.

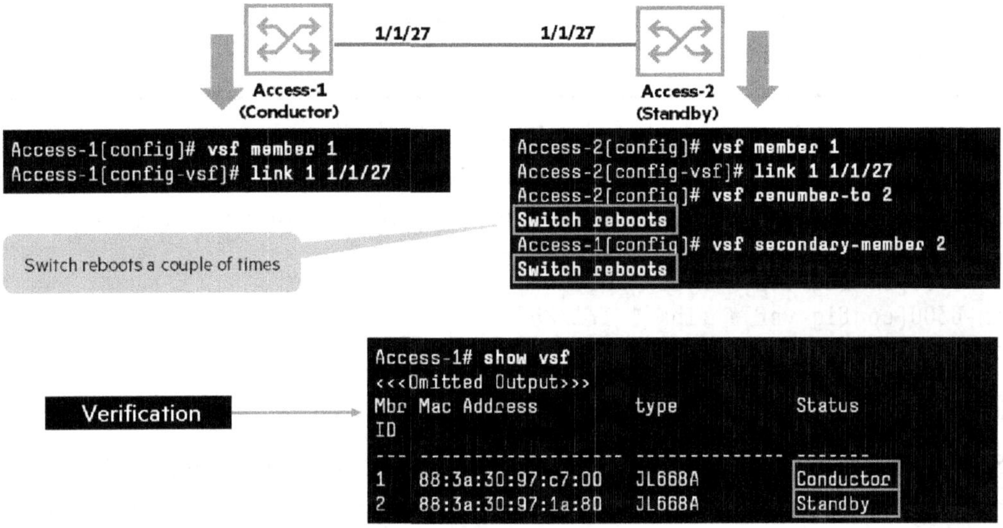

Figure 6-10: CX 6300 Series are assigned to the same VSF stack

Access-1 switch
```
Access-1[config]# vsf member 1
Access-1[config-vsf]# link 1 1/1/27
```

Access-2 switch
```
Access-2[config]# vsf member 1
Access-2[config-vsf]# link 1 1/1/27
Access-2[config]# vsf renumber-to 2
Access-2[config]#! Switch reboots
```

Adding a secondary member
```
Access-1[config]# vsf secondary-member 2
Access-1[config]#! Switch reboots
```

Use the `show vsf` command to determine member status and role.

Pre-provisioning

Pre-provisioning is an option to add new members to the stack, where the stack is previously configured to receive a new member, reducing the number of manual interactions and speeding up the process. To pre-provision the stack to receive a new switch, there are a few steps to be executed:

1. Select the member number for the new switch.
2. Select the switch model (part number) of the new switch.
3. Configure which ports should be used for the VSF links.

> The switch model (part number) must match the new switch model so it can join the stack automatically.

AOS-CX supports member pre-provisioning. This enables you to prepare the VSF link and member configuration for a specific CX 6200 Series or CX 6300 Series before the switch is connected to the stack. When the switch joins the stack, it boots up with the proper configuration.

```
Switch-6300(config-vsf)# vsf member 4
Switch-6300(config-vsf)# type jl658a
Switch-6300(config-vsf)# link 1 1/1/25
Switch-6300(config-vsf)# link 2 1/1/26
```

Auto-stacking

The VSF auto-stacking feature provides a mechanism to automatically form a stack when the stack members are physically connected in a desired topology. This reduces the number of user intervention touch points to form a VSF stack.

> It is important to note you must be running AOS-CX 10.07 or higher.

A manual stack formation procedure generally requires you to explicitly log in to each of the switch, configure the links, renumber the member ID, and then make the physical connection to form a VSF stack of desired size and topology. This is error prone since there are multiple touch-points involved in the whole work-flow for each member.

The auto-stacking feature eases this problem by reducing the number of touch-points involved to simple physical connections of the links. A new factory default switch can be added into an existing stack by physically connecting it to a given member of the stack. The new switch is automatically assigned a member ID, then it reboots. After reboot, the newly added member joins the stack.

There are two major components to the auto-stacking solution:

- **Peer discovery:** Initiated from the conductor using one of the methods of designating a conductor switch.
- **Auto-join eligibility:** Determined by the configuration state of each stack member. A switch with a factory default configuration is always auto-join eligible.

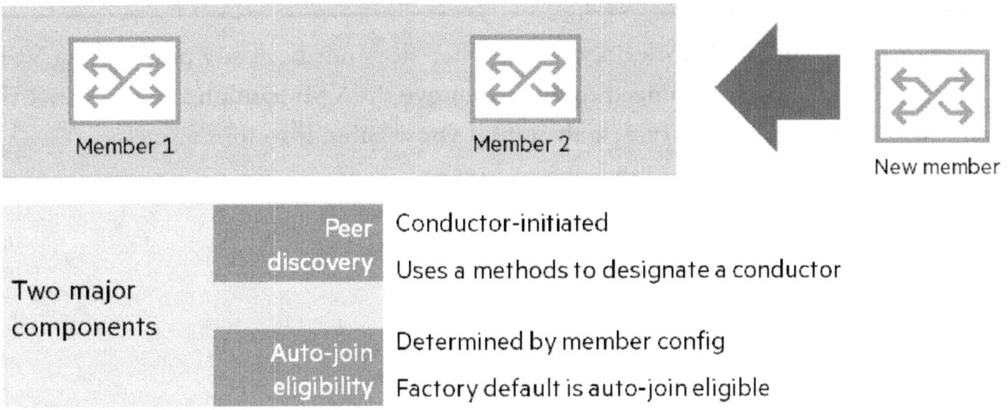

Figure 6-11: VSF auto-stacking

Peer discovery and auto-join eligibility

Peer discovery

Auto-stacking peer discovery is a unidirectional process. It starts on the conductor of the existing VSF with the VSF link containing the higher denomination VSF port. The conductor sends a VSF peer discovery protocol packet on that port. The peer receives the packet, determines if it is valid, and sends a response with information including its auto-join eligibility, MAC address, and part (model) type. If the peer is auto-join eligible, the VSF member and link configurations are automatically added to the running-configuration of the active conductor.

Auto-join eligibility

A switch in its factory defaults configuration state is considered to be auto-join eligible. If the auto-join eligible switch is connected to existing stack, it will automatically reboot and join the stack. Once it moves out of factory default configuration state, it is not considered as auto-join eligible and cannot automatically join an existing stack. However, an administrator can still manually configure the links, renumber the device to make it part of a stack.

Force auto-join support

Only a switch with a factory default configuration is considered to be auto-join eligible. In order to support factory express deployments where the administrator wants to add a switch which is in its non-factory default configuration, the force auto-join configuration support is provided. Use the `vsf force-auto-join` command to force the switch to join the stack automatically. Once the user sets force auto-join in the switch configuration, the switch will be considered as auto-join eligible and will join the stack even though the switch does not have the factory default configuration.

Force auto-join will work only if the switch does not have any pre-existing VSF configurations. You would need to either remove the VSF configuration or reset the switch back to its factory default state if you wanted it to join a stack.

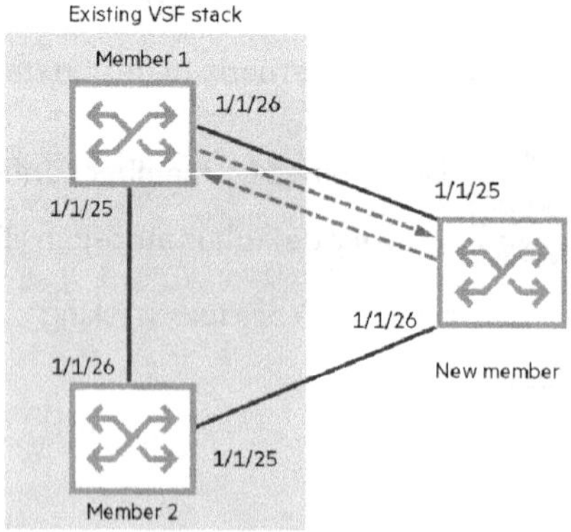

Figure 6-12: Peer discovery and auto-join eligibility.

Forming a VSF stack using auto-stacking command

1. Install the switches and physically connect them on the reserved auto-stacking interfaces.
2. On the conductor switch use the `vsf start-auto-stacking` command. The links and secondary member will be automatically configured.
3. Since the switches are already physically connected, each switch in the stack reboots automatically and join the stack one after another automatically.

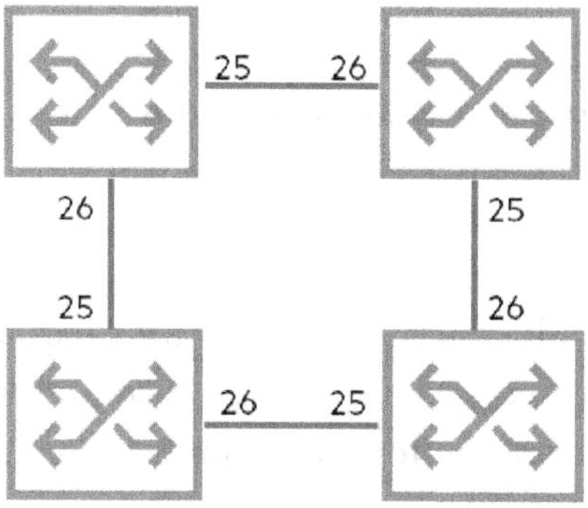

Figure 6-13: VSF stack using auto-stacking

Reserved interfacesfor auto-stacking

Based on the product type of a switch, the following two interfaces are reserved for the auto-stacking process:

- 24-port switch models: 25 and 26
- 48-port switch models: 49 and 50

Users can physically connect the switch to an existing stack on one of these reserved auto-stacking interfaces.

For auto-stacking, it is required that the existing stack conductor's VSF link port have a higher value. For example, if you connect member 2 to member 1 (existing VSF stack conductor), you must connect interface 26 of member 1 to interface 25 of member 2. Otherwise, the member 2 will not join in auto-stacking.

Table 6-1: Reserved interfaces for auto-stacking

Platform	SKU part number	Ports reserved
CX 6300 Series	JL658A JL660A JL662A JL664A JL666A JL668A	25 and 26
CX 6300 Series	JL659A JL661A JL663A JL665A JL667A JL762A	49 and 50
CX 6200 Series	JL724A JL725A	25 and 26
CX 6200 Series	JL726A JL727A JL728A	49 and 50

Member ID and port numbers

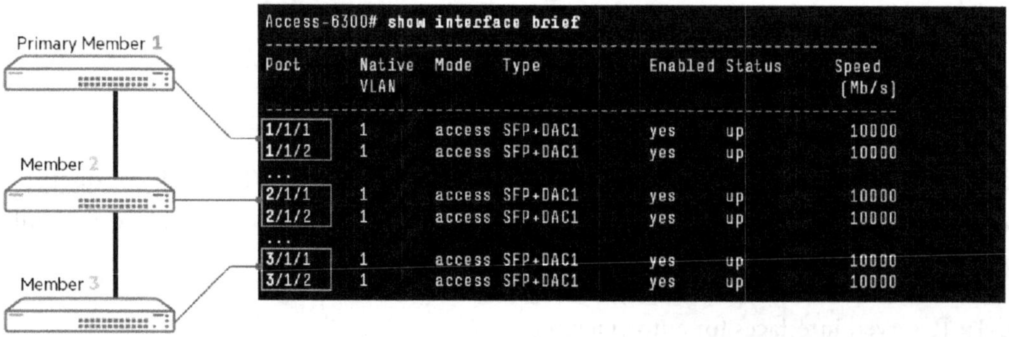


```
Access-6300# show interface brief
---------------------------------------------------------------------------
Port       Native  Mode    Type            Enabled Status         Speed
           VLAN                                                   (Mb/s)
---------------------------------------------------------------------------
1/1/1      1       access  SFP+DAC1        yes     up             10000
1/1/2      1       access  SFP+DAC1        yes     up             10000
...
2/1/1      1       access  SFP+DAC1        yes     up             10000
2/1/2      1       access  SFP+DAC1        yes     up             10000
...
3/1/1      1       access  SFP+DAC1        yes     up             10000
3/1/2      1       access  SFP+DAC1        yes     up             10000
```

When the VSF stack forms, all physical devices use a single control plane. This means that all switch interfaces are available for configuration, using the standard AOS-CX Member/Slot/Port notation. Use the `show interface brief` command to see all available ports for all switches in the stack.

> Since the CX 6200 Series and CX 6300 Series models are fixed switches, it is a 1U switch with only one slot. The slot number is always 1.

Tracing Layer 2 traffic: unicast

VSF stacked switches operate as a single virtual switch, as VSF combines the management, control, and data planes. Therefore, the traffic forwarding is similar to the traffic forwarding in a single switch, with the exception that the traffic may arrive in one switch of the stack with the destination to another switch in the same stack. This means that the switch that receives the traffic will need to transmit the packet over the VSF link to the destination switch, to deliver it to its destination.

VSF traffic delivery process:

- Each member uses the forwarding table to look for the destination address location.
- If the destination is on a different member, it will use the VSF link to transmit the packet.
- User traffic can traverse other members.
- When the destination egress equals the local switch, the traffic will be sent out to the destination port.

```
Switch# show mac-address
MAC Address          VLAN     Type          Port
-----------------------------------------------------
90:20:c2:bc:ee:00    10       dynamic       2/1/1
00:0b:86:b4:eb:37    10       dynamic       3/1/1
```

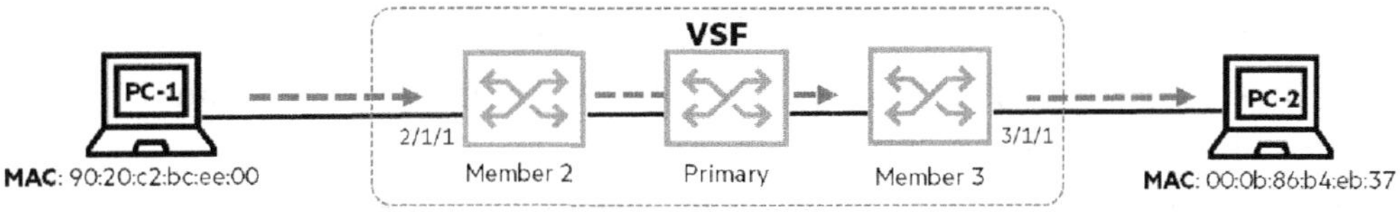

Figure 6-14: Tracing Layer 2 traffic unicast

A VSF fabric is like any switch configured with link aggregation. It learns MAC addresses on a logical LAG entity rather than the physical interfaces. It selects one LAG member link to forward each conversation.

However, VSF overrides the typical LAG hash function used for physical interface selection. A VSF member prefers to use its own local links and avoids using VSF links. If the member has multiple local links in the aggregation, then it uses the typical hashing mechanism to choose between those.

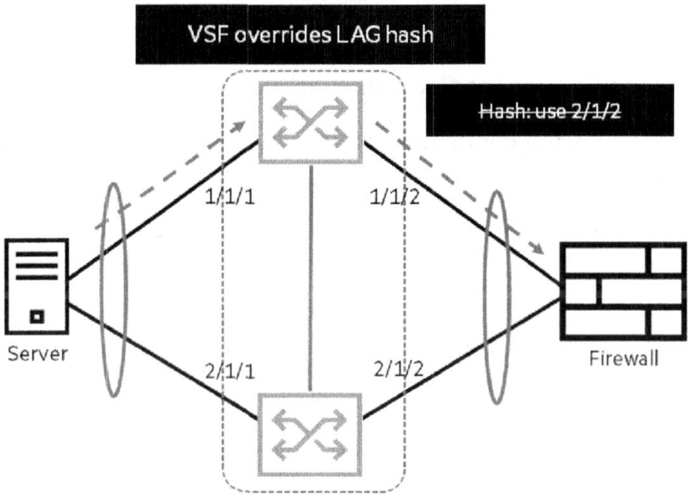

Figure 6-15: VSF overrides LAG hash

Failover and OSPF graceful restart

A functional stack must be configured with a standby for redundancy purposes. If the conductor fails and there is no standby, the stack will fail.

When the conductor device fails and a standby device is in place, the standby seamlessly assumes the role of the new conductor, ensuring uninterrupted stack management. Once the old conductor is replaced, it seamlessly assumes the standby role, providing a smooth transition.

When the primary (conductor) switch fails, the standby switch will assume the conductor role. That change may affect the OSPF neighbors list and temporarily disrupt the network while the new switch establishes the OSPF neighborhood and learns new routes. To avoid that problem, OSPF graceful restart may be used. With OSPF graceful restart, routing can be gracefully restarted on the switch without losing packets that are in transit. OSPF neighbors are informed that the router is completing a graceful restart, which allows for maintenance on the switch without interrupting traffic in the network.

> Graceful restart is only applicable to the CX 6200 Series and CX 6300 Series VSF switches.

Procedure

1. Manually create a router ID configuration to support graceful restart and management module failover. A dynamic router-id is chosen from the list of configured IPs at the time of the router-id election. In the event of a graceful restart and management module fail-over, the router-id should be preserved before and after these events. Graceful restart would fail if there is a change in the router-id, causing traffic loss.
2. Enable graceful restart of OSPF routing using the `graceful-restart {restart-interval <seconds> | helper}` command.

   For example, the following command specifies 50 seconds as the maximum interval another router will wait for this router to gracefully restart.

   ```
   switch(config-ospf-1)# graceful-restart restart-interval 50
   ```

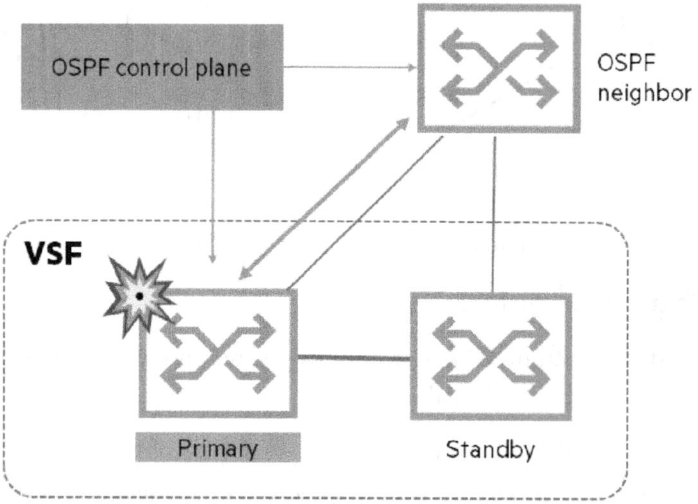

Figure 6-16: Failover and graceful restart where conductor device fails

What happens with VSF coordinator failure?

1. OSPF peers are notified.
2. A new conductor runs the OSPF process.
3. Traffic is forwarded during convergence.

The secondary assumes the conductor role upon primary failure. The new conductor runs all control plane protocols, uses the configuration databases, and responds to management sessions.

When the primary member fails, Layer 2 traffic continues to be forwarded without disruption. Layer 3 protocols, such as OSPF, perform graceful restart by notifying peers of the failover event. This triggers a rebuild of OSPF adjacencies. During convergence, the switch continues to route traffic based on the last-known routing information prior to the failure. This typically lasts only a few seconds. After OSPF is fully operational, routing uses the new information.

With VSF, there is no preemption. This means that when the failed member re-joins the stack, it will not replace the current conductor of the stack. Instead, it takes the standby role.

Link failure

VSF link failures could cause a fragmented stack.

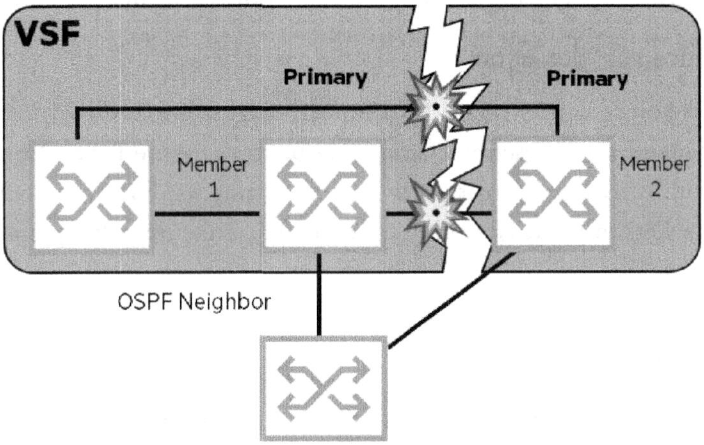

Figure 6-17: VSF link failure topology

The figure shows a fragmented VSF stack, due to link failures. Originally, SW1 was primary and SW2 was secondary. After fragmentation, there is no direct connection between the two switches. SW1 continues to think that it is the primary. SW2 no longer hears from SW1 and so now thinks that it too is the primary. This is known as a split-brain condition—both fragments continue to function although there is no communication between them.

Split-brain can cause unstable, unexpected network behavior. A packet received in one fragment and destined for the other fragment is discarded. Even worse, both segments use the same IP address and the same routing information; external devices could start populating duplicated data. This can cause strange network behavior.

A split-brain situation could also occur if a VSF member fails in a daisy chain topology.

The best way to solve a split-brain situation is to disable the ports in one of the segments.

Multi-active detection - split detection

You can use Multi-Active Detection (MAD) to avoid split-brain situations. Thus, if a VSF link failure occurs, the segment that includes the standby member verifies the primary member status. If the original primary is up, then all members in the fragment that does not include the primary member will disable all their ports.

VSF uses two mechanisms to detect and verify the status of the primary member.

Management interface split detection

This method requires you to connect Out-Of-Band Management (OOBM) interfaces to primary and secondary stack members. These interfaces must be in the same Layer 2 broadcast domain (VLAN). This network is used to identify active stack fragments. Each member broadcasts split detection protocol packets to identify stack fragments that are currently operational.

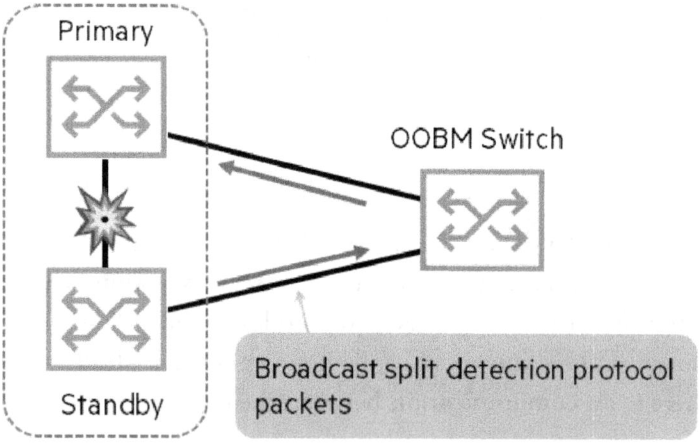

Figure 6-18: Multi-active detection—split detection

HPE Aruba Networking Virtual Switching eXtension

HPE Aruba Networking Virtual Switching eXtension (VSX) is a cutting-edge technology that revolutionizes network infrastructure by enabling the creation of a virtual chassis from multiple physical switches. With VSX, organizations can achieve unparalleled levels of redundancy, scalability, and manageability in their network environments, empowering them to meet the demands of modern network switch challenges with ease.

VSX overview

VSX is a virtualization technology for AOS-CX switches. VSX is commonly implemented in core devices and at data centers. VSF is suitable for the access layer.

VSX improves data plane performance. With VSF, the control plane can be run only in the primary member. Thus, some time is wasted when non-primary members ask the control plane how to handle packets. With VSX, each member runs its own control plane, allowing for faster decisions, reduced latency, and better performance.

Although VSX switches run separate control planes, they still maintain database synchronization for the configuration. Unlike VSF, each switch can modify and populate the control plane, while presenting themselves as one virtualized switch to other devices.

VSX allows you to upgrade members with near zero downtime and with continuous packet forwarding.

This technology can be run in the CX 6400 Series, CX 8100 Series, CX 83xx Series, CX 8400 Series, CX 9300 Series, and CX 10000 Series. The CX 6200 Series and CX 6300 Series only run VSF.

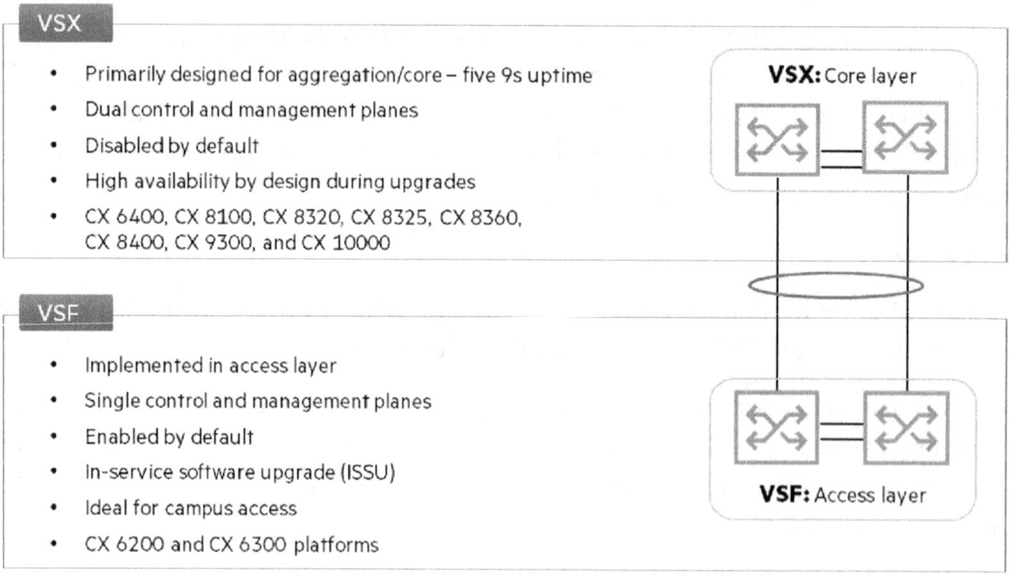

Figure 6-19: VSX core layer and VSF access layer

Switch virtualization solution comparison

VSX virtualizes the control plane of two switches to function as one device at Layer 2 and as independent devices at Layer 3. From a datapath perspective, each device does an independent forwarding lookup to decide how to handle the traffic. Some of the forwarding databases, such as the MAC forwarding database and neighbor tables, are synchronized between the two devices using a proprietary VSX control plane. Some of the forwarding databases are built independently by each switch. Configuration synchronization is one aspect of this VSX solution where the primary switch configuration is synced to the secondary switch.

VSF allows multiple ArubaOS switches of the same model to act as a single virtual device. VSF helps to provide fast failover, scalability, manageability, and high availability.

A VSF fabric has one active management plane and one active control plane, which the management module in the commander switch provides. The VSF commander proxies the control plane to the standby member. Both members of the fabric participate in the forwarding plane. Interface modules from both members are combined as if over one large switch connected by the VSF link.

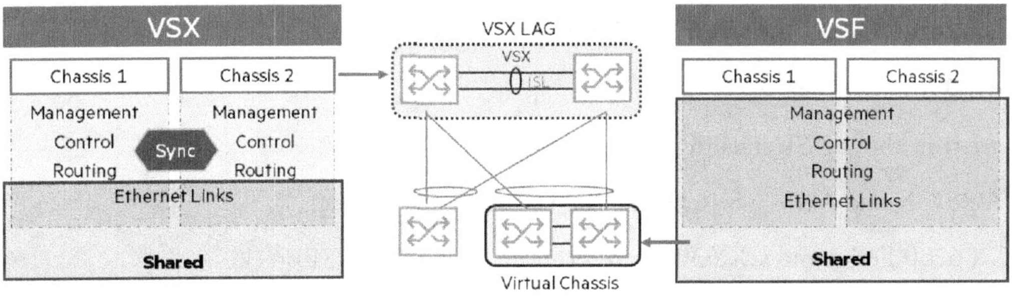

Figure 6-20: VSX and VSF comparison

VSF versus VSX

The figure summarizes the differences between VSF and VSX, as a study and review aid.

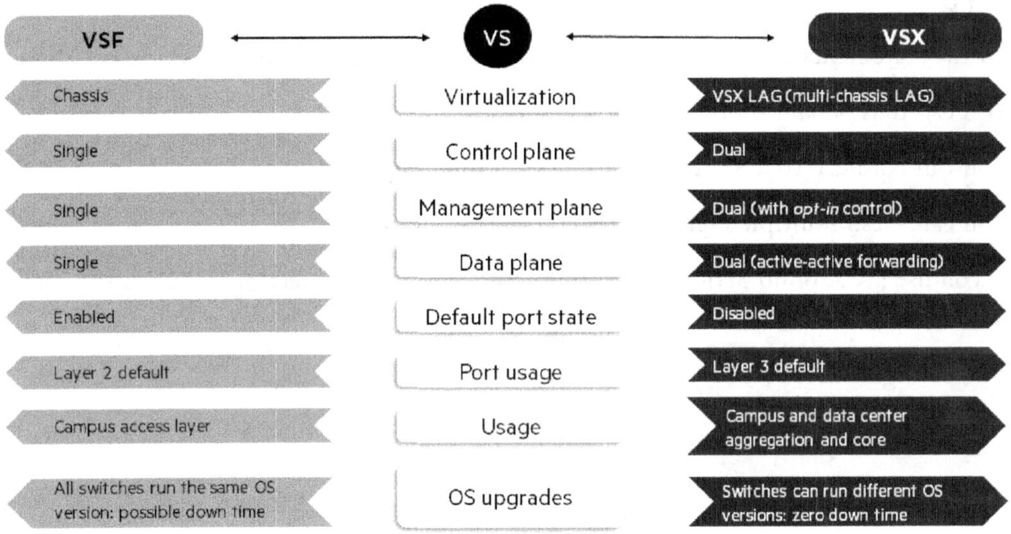

Figure 6-21: VSF versus VSX

Learning check

Check your understanding of this material by answering the following questions.

Question 1

Which of the statements below accurately describe network devices' operational planes? (Select two)

- ☐ The control plane is used to control the switch configuration.
- ☐ The management plane manages internal Layer 2 and Layer 3 switch processes.
- ☐ The data plane moves data from ingress to egress port.

- ☐ The control and data planes are tightly integrated.

Question 2

Which are true about VSF stacking? (Select two)

- ☐ AOS-CX and AOS-S switches can be mixed in a VSF stack.
- ☐ CX 6200 Series and CX 6300 Series can be mixed in a VSF stack.
- ☐ Different models of CX 6200 Series can be mixed in a VSF Stack
- ☐ CX 8325 Series VSF stack supports up to 10 switches.
- ☐ The use of VSF is recommend in the access layer of a campus network.

Question 3

Which of the statements describes valid VSF requirements and specifications? (Select three)

- ☐ You can daisy-chain up to 10 VSF members.
- ☐ You can connect 10 VSF members in a ring topology.
- ☐ You can mesh multiple members together for redundancy.
- ☐ If you use a CX 6300 Series as the conductor, you can connect 3810 Series switches as members.
- ☐ The configuration of VSF may cause members to reboot.

7 Layer 3 routing

Welcome to the Layer 3 routing chapter, where we will discuss the foundational concepts of routing in computer networks. In this chapter, we will explore the intricacies of how routers forward data packets between different networks, enabling communication across diverse network environments. From understanding IP addressing and subnetting, to exploring default gateways, inter-VLAN and Virtual Routing and Forwarding, this chapter will help set the stage of Layer 3 routing.

OBJECTIVES

After completing this chapter, you should be able to:

- ✓ Define Layer 3 routing and its significance in forwarding data packets between different networks.
- ✓ Describe inter-VLAN routing and the significance in network segmentation.
- ✓ Demonstrate virtual routing and forwarding and its operational relevance in network architecture.

Routing introduction

Routing refers to the process of directing data packets from their source to their destination across a network, making dynamic decisions based on network topology, available paths, and routing algorithms. It involves making the most efficient path for data transmission, where the destination address is examined and appropriate routes are selected. Routing enables effective communication between devices across interconnected networks, facilitating the exchange of information.

Layer 2 forwarding versus Layer 3 routing

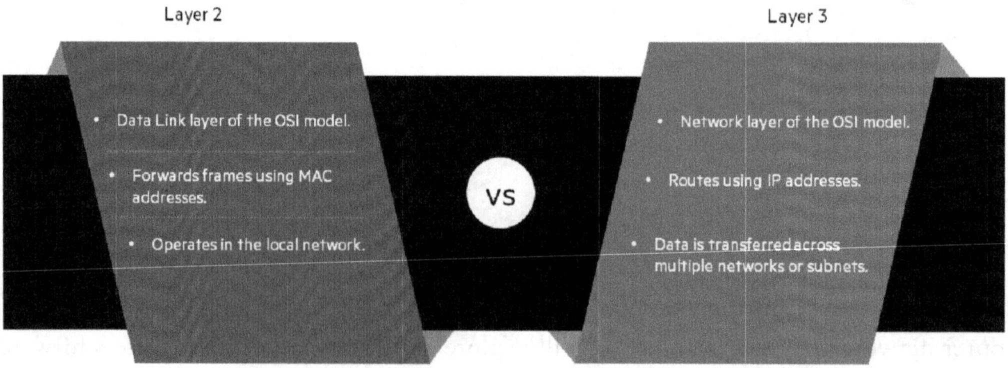

Figure 7-1: Layer 2 versus Layer 3 routing comparison

What is the difference between Layer 2 and Layer 3 routing? These layers refer to the different levels of the OSI model and serve distinct functions in networking. Layer 2 routing operates at the data link layer of the OSI model and primarily involves forwarding packets based on MAC addresses. Switches make these forwarding decisions using MAC address tables and operate within the same local network or LAN (Local Area Network).

Layer 3 routing operates at the network layer of the OSI model and involves forwarding packets based on IP addresses. Routers make routing decisions using routing tables, which contain information about network destinations and paths. Layer 3 routing enables communication between different networks or subnets, allowing data to traverse across multiple networks to reach its destination.

Layer 3 routing

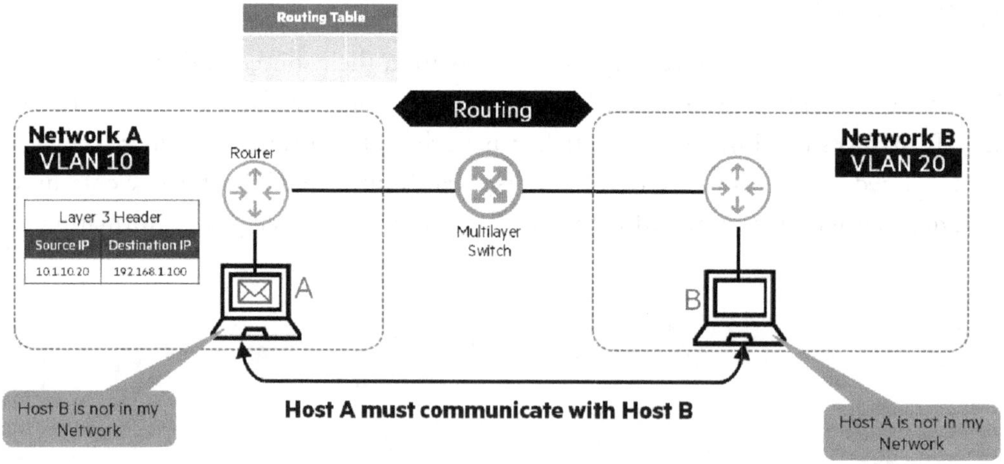

Figure 7-2: Layer 3 routing of data packets

Here is a simplified explanation of how data packets are forwarded at Layer 3.

1. **Destination address lookup:** When a router receives a data packet, it examines the destination IP address contained within the header.
2. **Routing table consultation:** The router consults its routing table, which contains information about known network destinations and associated routes. This table is populated either through static configuration or dynamically through routing protocols.
3. **Best path determination:** Using the destination IP address, the router selects the best path or next-hop router to reach the destination, making process decisions such as the shortest path, lowest latency, and available bandwidth.
4. **Packet forwarding:** Once the best path is determined, the router forwards the data packet to the appropriate outgoing interface or next-hop router along the selected route.
5. **Repeat process:** If the destination network is not directly connected, the process repeats as the packet is forwarded from router to router until it reaches its final destination.

Overall, network layer routing involves intelligent forwarding of data packets across interconnected networks based on IP addressing and routing information.

Request for Comment (RFC) 1918 defines the address allocation for private networks. The Internet Assigned Numbers Authority (IANA) has reserved the following three blocks of the IP address space for private internets:

- 10.0.0.0 - 10.255.255.255 (10/8 prefix)
- 172.16.0.0 - 172.31.255.255 (172.16/12 prefix)
- 192.168.0.0 - 192.168.255.255 (192.168.16 prefix)

As those address blocks should not be advertised on the internet, NAT (Network Address Translation) is needed to communicate those networks to the internet.

Layer 3 analysis

Layer 3 analysis involves examining various aspects of routing at the network layer. The analysis includes the following:

- **Routing protocols**: Evaluating the operation and effectiveness of routing protocols such as RIP, OSPF, EIGRP, and BGP in dynamically exchanging routing information and building routing tables.
- **Routing table analysis**: Studying routing tables maintained by routers to understand how routing decisions are made based on destination IP addresses and associated routes.

- **Routing metrics**: Analyzing the metrics used by routing protocols to determine the best path for packet forwarding, including factors like hop count, bandwidth, delay, and path cost.
- **Routing convergence**: Assessing the speed and efficiency at which routers converge on consistent routing information in response to network change, ensuring stable and optimal routing paths.
- **Route redistribution**: Examining the process of redistributing routes between different routing domains or protocols, ensuring seamless communication between disparate parts of the network.
- **Routing policies**: Reviewing routing policies and configurations to ensure alignment with network objectives and requirements, including traffic engineering, security policies, and Quality of Service (QoS) parameters.

Overall, Layer 3 routing analysis aims to optimize network performance, scalability, and reliability by understanding and optimizing the routing processes that enable communication between devices and networks in complex network environments.

Loopback interfaces

A loopback interface is a virtual interface supporting IPv4/IPv6 address configuration. The loopback interface can be considered stable because once created, it remains up. The loopback interface can then be configured with an address to use as a reference or identifier independent of the physical interfaces.

Device identification:

As long as the router is operational, the state of the loopback interface is always up. Even if only one link to the router is active, the loopback interface can be reached. This functionality makes it possible to identify an active device in the network using the IP address configured on the loopback interface.

Routing:

Since the loopback interface is always active, a routing session (such as a BGP session) can continue on an alternate path even if the outbound interface fails. In OSPF, a loopback interface address is advertised as an interface route into the network. This functionality increases reliability by allowing traffic to take alternate paths if there is a link failure. In OSPF and BGP, the router ID can be set to the loopback address to avoid reassignment of the router ID when physical interfaces are added or removed.

Device management:

Loopback interface is always reachable and can be used for sending and receiving management information such as logs and SNMP traps without interruption.

Figure 7-3: Loopback interfaces

IP routes and default gateway

The default gateway plays a critical role in IP routing by serving as the exit point for data packets that need to reach destinations outside the local network. When a device attempts to communicate with a destination on a different network, it sends the data packet to the default gateway. The default gateway then forwards the packet to the final destination, either directly or through other routers, based on the routing table entries. Essentially, the default gateway acts as the router of last resort, enabling devices within a network to communicate with external networks and the internet.

IP routes

When a device must communicate with others in a different network, it must know which local network device on its broadcast domain can route the traffic toward the destination network. This information is provided to computers using IP routes.

For endpoints, this information can be manually added, in the form of a so-called "static route." Endpoints can also have this assigned to them through DHCP in the form of the "default gateway." Routers and multilayer switches can use manually added static routes, or they can dynamically and automatically determine the best routes to each destination using a routing discovery protocol. You will soon learn about dynamic routing discovery protocols in a later chapter.

A static route must specify the following information:

- **Destination IP address:** Where you want to go.
- **Subnet mask:** How that destination IP address is split into a network portion and a host portion.

- **Next hop IP address:** The IP address of the router on your network that can route packets to the destination.

The scenario in the following figure shows three networks connected by two routers. Router-1 connects Networks A and B, while Router-2 can route between Networks B and C.

If Host A (in Network B) must communicate with Server-1 in Network A, it must use the first router entry. The first entry states: "To get to destination 10.0.0.1, which has a mask of 255.0.0.0 (/8), you must use the next-hop router at IP address 172.16.0.1."

What if Host A must use Server-2? The second route entry states: "To get to destination 192.168.0.1, which uses a 255.255.255.0 (/24) mask, you must send the packets to the next-hop router at IP address 172.16.0.2."

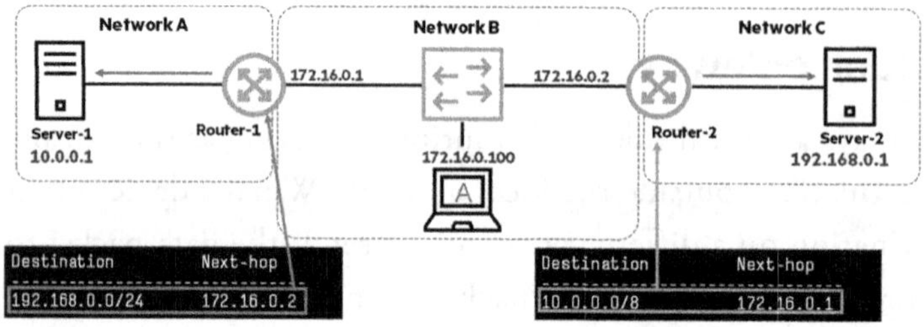

Figure 7-4: IP route of three networks connected by two routers.

A host can manually add routes into its table. This solution does not scale.

Default gateway

A default gateway is sometimes referred to as the "gateway of last resort." Most host devices only need to be configured with a default gateway address, since they often only have one router on their network. Routers and switches, however, will often have multiple entries in their route tables, providing different paths to get to different networks. If none of these route entries define a path to the desired destination network, then the default gateway will be used. The default gateway's core objective is to create a unique route that summarizes all possible destinations, which optimizes host routing information.

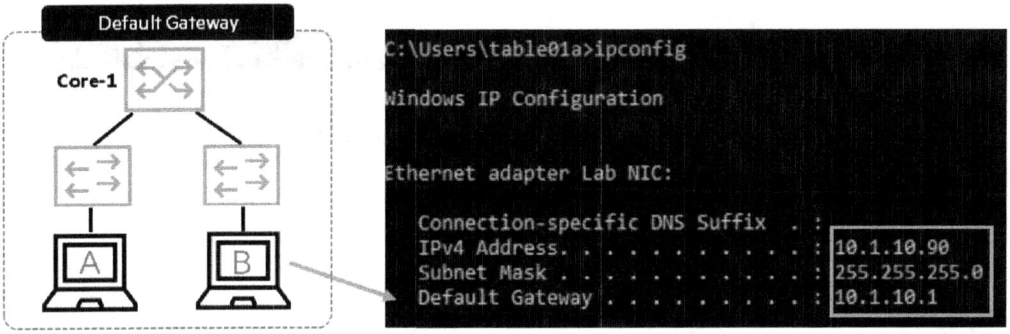

Figure 7-5: Router Core-1 acting as the default gateway for all devices in Network A or B

The default gateway address specified on a device (such as laptop B in the figure) defines the router that laptop B will use when forwarding any packet to another network. It is like telling a host, "To get to everywhere in the known universe, go to this next-hop address." The default gateway optimizes and simplifies endpoint routing decisions, since only a single route is typically required for any endpoint device.

> Endpoints typically have only one default gateway.

Inter-VLAN routing

Inter-VLAN routing is the process of enabling communication between different VLANS within a network environment. This is typically accomplished using a router or Layer 3 switch capable of routing traffic between VLANs. By facilitating communication between VLANs, inter-VLAN routing allows devices in different VLANs to exchange data and access resources located on separate VLANs, enhancing network segmentation and security while maintaining connectivity.

Introduction to multilayer switch

In years past, multilayer switches did not exist. Older environments used Layer 2 switches for host connectivity and then routed between them with an external router. You see this in the figure's left-hand example. A potential problem with this deployment is that the switch-to-router link can become oversubscribed, although the LAG can alleviate this problem to some extent. Performance can also be suboptimal because sending frames to the router requires an additional routing decision.

Multilayer switches are more efficient devices. The switching and routing functions of the device are connected via a high-speed internal backplane. Initial routing decisions and other processes all happen "in the box." This can reduce latency and increase performance.

All AOS-CX switches are multilayer switches, which have routing enabled by default.

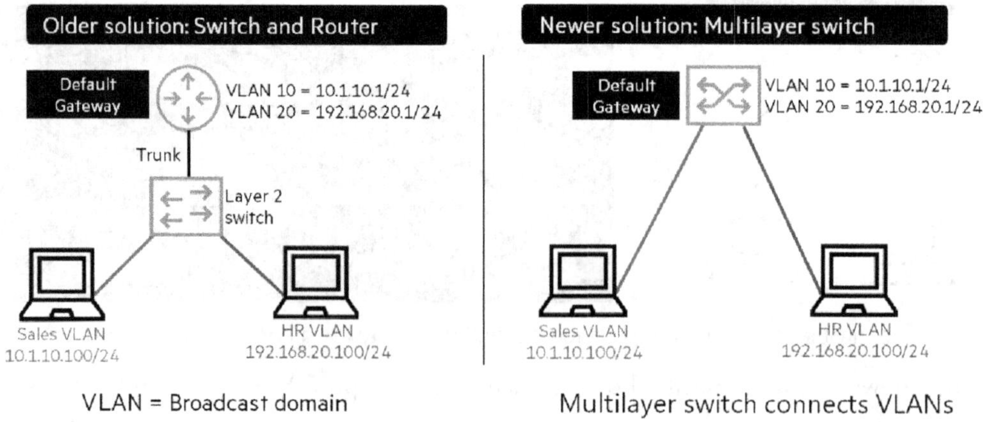

Figure 7-6: Switch router versus multilayer switch

Multilayer interface types

Multilayer interface types refer to network interfaces on devices such as routers or Layer 3 switches that have the capability to operate at multiple layers of the OSI model. These interfaces can process traffic at both Layer 2 and Layer 3, enabling them to perform functions such as routing and switching simultaneously. Multilayer interfaces are commonly used in scenarios where flexibility and efficiency are required, allowing devices to handle both local traffic switching at Layer 2 and inter-network routing at Layer 3 within the same interface.

A switch port can only process Layer 2 features and protocols—like being a member of a VLAN, for example.

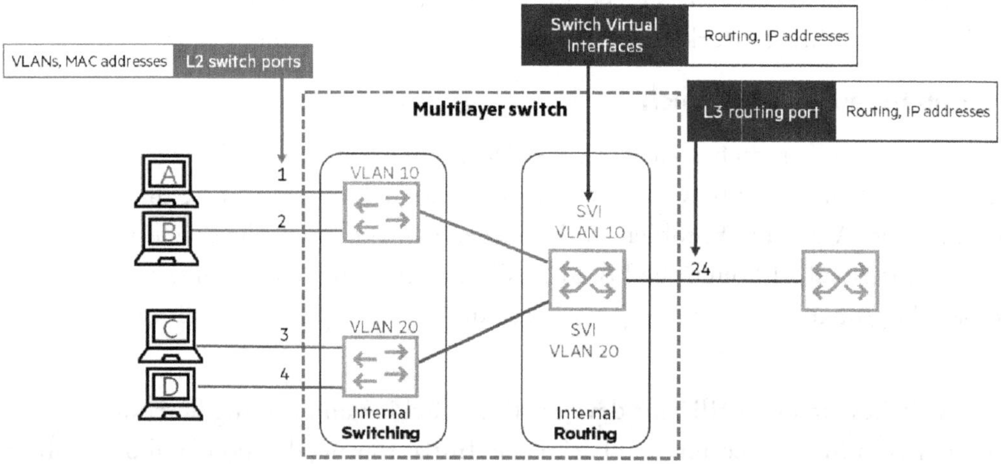

Figure 7-7: Ports 1-4 being used as Layer 2 interfaces

Ports 1-4 attach to end systems, accept Layer 2 frames as being a member of some Layer 2 VLANs, and then forward them to their destination Layer 2 MAC address. You can enable or disable Layer 3 features on an individual port by issuing the `no routing` command (for Layer 2 support) or the "routing" command (for Layer 3 support).

The AOS-CX switches are multilayer switches—they have both internal Layer 2 switching functions and internal Layer 3 routing capabilities. Because of this, you can route between VLANs. You need to connect each Layer 2 VLAN to the internal routing functions. To do this, create Switch Virtual Interfaces (SVI).

Suppose that you define SVI 10. Because it is an SVI, by definition it connects to the internal routing construct. Because it is SVI "10," by definition it connects to VLAN 10 and services routable traffic from VLAN 10 to other destination networks.

Similarly, you might define SVI 20. With some routing configuration, your switch can route traffic between your VLANs.

Next, consider that you need to connect your multilayer switch to an external router, perhaps using port 24, as shown in the figure. Since all ports are Layer 3 interfaces by default, port 24 connects to the internal routing functions by default; you merely need to configure it with typical Layer 3 parameters, such as an IP address. You will soon learn about these concepts and syntax.

The SVIs are virtual Layer 3 interfaces, for internal routing and port 24 is a physical Layer 3 interface for external routing. All are Layer 3 interfaces, and so perform routing functions. They accept routable Layer 3 packets and forward them based on their destination IP address.

DHCP helper address

A Dynamic Host Configuration Protocol (DHCP) helper address is a configuration setting typically found on routers or Layer 3 switches in a network environment. Its primary function is to facilitate the distribution of IP addresses and other network configuration parameters to client devices that are located on different subnets or VLANs.

When a client device on a subnet sends a DHCP request broadcast to obtain an IP address lease, the DHCP helper address feature intercepts this broadcast and forwards it as a unicast message to a designated DHCP server located on a different subnet. This allows the DHCP server to receive the request and respond with the appropriate IP address and configuration details, even though the client and server are not directly connected on the same subnet.

Commonly, hosts will automatically get an IP address by using the Dynamic Host Configuration Protocol (DHCP).

Endpoints broadcast a DHCP request: "Hey everyone, I need an IP address, a mask, and a default gateway." Because it is a broadcast, the host and server must be on the same subnet. Remember, a router defines the edge of a broadcast domain and does not forward a broadcast. Routers on different broadcast domains (VLAN) do not hear the request, and so no address is assigned.

If an organization deploys thousands of broadcast domains (VLANs), then you would need thousands of DHCP servers—one per VLAN! This is not realistic. You need a central DHCP service for all VLANs.

The solution is to configure a DHCP helper address on each router interface that serves as the default gateway for endpoints. The following process describes how the solution works when a router is properly configured with a helper address:

1. The client broadcasts a normal DHCP query.
2. The router that is on the client's network (the client's default gateway) receives this broadcasted DHCP query.
3. Instead of discarding the broadcast, as is normal, the router "helps" this broadcast by forwarding it on to the DHCP server. It converts this broadcast into a unicast, with the destination address specified in the `ip helper-address` command. Now that the message is a unicast, the router forwards it as it would any other unicast packet, toward its final destination—the DHCP server.
4. The DHCP server receives the DHCP request and replies with a DHCP offer—a unicast message sent to the requesting host, via the router.

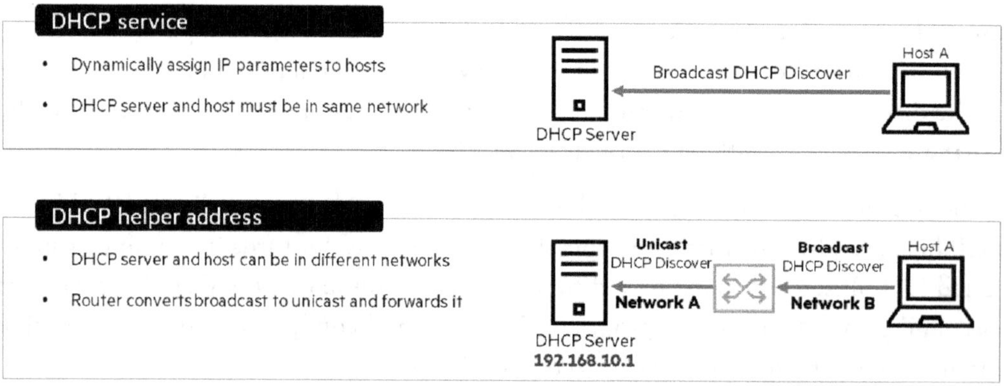

Figure 7-8: DHCP helper address sends unicast packet

In AOS-CX, the `ip helper-address` command defines the address of a remote DHCP server. Up to eight addresses can be defined. When more than one DHCP server has been defined, the switch will send the client request to all defined servers.

The following command has to be used under a Layer 3 interface:

```
Switch(config)# interface vlan 1
Switch(config-if)# ip helper-address <IP-ADDRESS>
```

IP routing table

An IP routing table is a data structure maintained by routers in a network environment that contains information about known network destinations and the best paths or routes to reach them. This table is a critical component of the routing process and guides routers in making decisions about how to forward data packets to their intended destinations.

Each entry in the routing table typically includes the following information:

1. **Destination Network:** The network address of the IP prefix of the destination network.
2. **Next-hop Router**: The IP address of the next-hop router or exit interface that should be used to reach the destination network.
3. **Interface**: The outgoing network interface through which packets should be forwarded to reach the next-hop router or directly connected network.
4. **Metric**: An optional value representing the cost or preference associated with the route, used by routers to determine the best path among multiple routes to the same destination.

Routing tables are populated through various mechanisms, including manual configuration, dynamic routing protocols, and default routes. Routers use the information in the routing table to make forwarding decisions for incoming data packets, selecting the best route based on the destination IP address and associated routing table entries.

Periodically, routing tables are updated to reflect changes in network topology, link status, or routing information received from neighboring routers. This ensures that routers have accurate and up-to-date information for making routing decisions, facilitating efficient and reliable data packet forwarding across the network.

Routing devices (routers and multilayer switches) build and maintain a routing table that informs them of the best path to any given destination.

You can manually add entries to the route table, in the form of static routes. Alternatively, you can configure a routing protocol, which automatically builds and maintains this table. Finally, you can have a default gateway defined. Typically, entries in the routing table do not expire unless a change in the topology causes an update. This differs from the MAC address table in Layer 2 switches where an entry expires after five minutes if the switch stops receiving traffic from the endpoint.

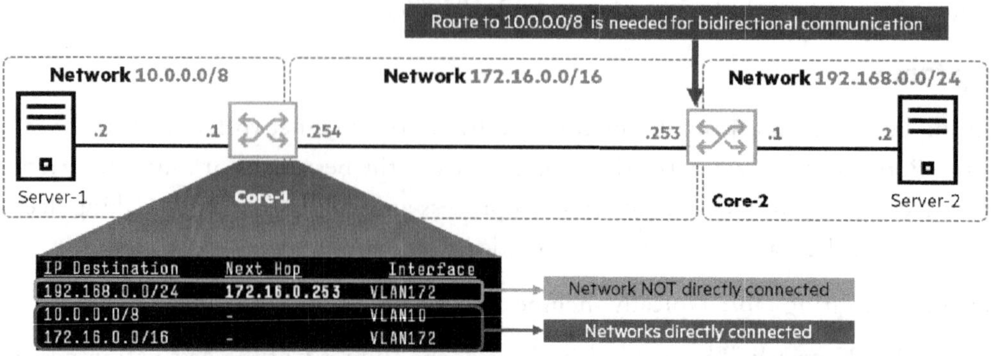

Figure 7-9: A simplified view of the route table on the Core-1 router.

There are three networks: 10.0.0.0/8, 172.16.0.0/16, and 192.168.0.0/24. Recall that the subnet mask determines the network portion of an IP address.

Analyze Core-1's route table. The first entry states, "To get to any host on network 192.168.0.0/24, send packets to next-hop router 172.16.0.253 (Core-2). To get to that next-hop address, forward the packet out local interface VLAN172."

The next entry says, "To get to any host on network 10.0.0.0/8, there is no next-hop...I'm directly connected to that network. Simply forward the packet out my local VLAN interface 10." Finally, to connect to network 172.16.0.0/16, Core-1 is directly connected to this network so there is no need to assign a next-hop; simply forward the packet out of local VLAN interface 172.

Consider, at this moment, that you configure only Core-1; when Server-1 connects to Server-2, Core-1 will properly route the packet and send it to Core-2. This device will have no problem sending the packet to Server-2; this network is locally connected to Core-2. So, one-way communication is successful. However, the response from Server-2 is sent to Core-2; this device receives the packet, but since the destination address of Server-1 (10.0.0.2) is not in its route table, the packet will be dropped. Simply put, bidirectional communication cannot be done. To solve the problem, Core-2 will need to be configured with a route to reach the non-directly connected network (10.0.0.0/8).

Packet delivery

Packet delivery is the cornerstone of network communication, encompassing the process of transmitting data packets from a source device to a destination across a network infrastructure. It involves a series of steps, including packet creation, addressing, routing, and eventual delivery to the intended recipient. Understanding the intricacies of packet delivery is essential for optimizing network performance, ensuring data integrity, and facilitating seamless communication between devices in diverse network environments.

Scenario

In this scenario, PC-1 must communicate with PC-2. Both endpoints connect to the same Layer 2 switch, but they are mapped to different VLANs. The Core-1 multilayer switch is there to do inter-VLAN routing.

- PC-1 has IP address 10.1.10.100/24 on VLAN 10, and its DG is 10.1.10.1. It connects to port 1 of Access-1.
- PC-2 has IP address 10.1.20.100/24 on VLAN 20, with DG = 10.1.20.1. It connects to port 2 of Access-1.

Access-1 port 24 connects to Core-1 port 23. This connection is a trunk connection allowing VLANs 1, 10, and 20, with VLAN 1 being the native VLAN.

Collapsing Layer 2 access services and Layer 3 routing services into a single multilayer switch can improve efficiency. However, for larger networks, there is typically a separate layer of pure Layer 2 switches for endpoint access, connected to a smaller set of Layer 2/Layer 3 multilayer switches. This design improves scalability.

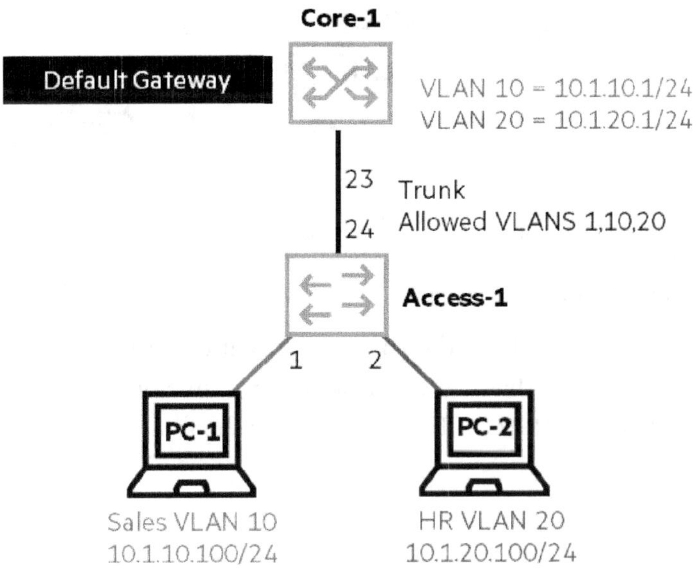

Figure 7-10: PC-1 must communicate with PC-2, but they are mapped to different VLANs

Endpoint to access switch

PC-1 generates a message that contains the following information:

- **Layer 3 header:** The source IP address is PC-1's 10.1.10.100 and the destination IP is PC-2's 10.1.20.100.
- **Layer 2 header:** The source MAC address is PC-1's MAC address and the destination MAC address is the MAC address of the default gateway on VLAN 10 (the MAC address associated with 10.1.10.1).

Remember, if PC-1 does not know the MAC address for 10.1.10.1, it performs an ARP process to get this information.

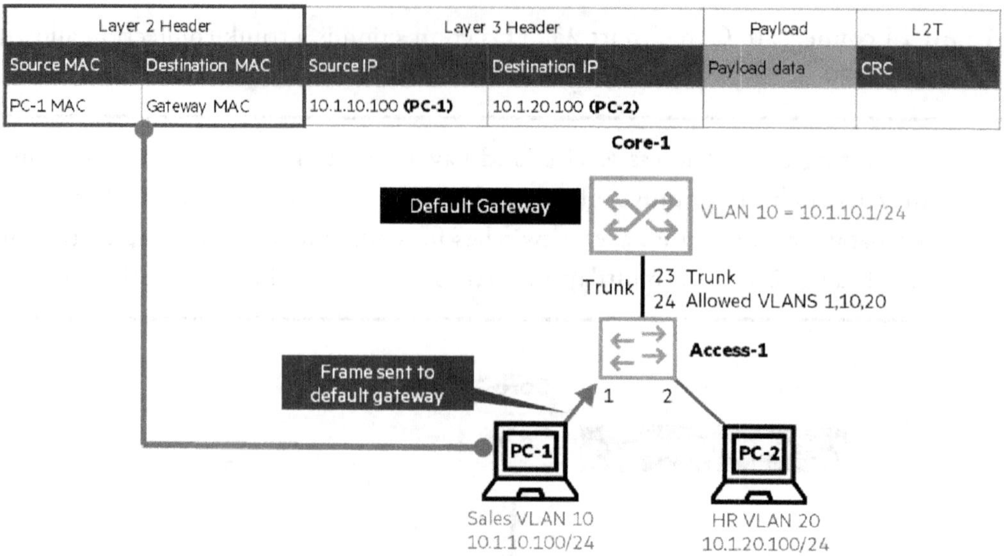

Figure 7-11: Endpoint to access switch

Access to multilayer switch

Access-1 receives the frame and analyzes the Layer 2 destination MAC address. It finds a match in its MAC address table and knows that it must forward this frame out its trunk link, port 24, to Core-1. It adds an 802.1Q tag, VLAN = 10, and forwards the frame.

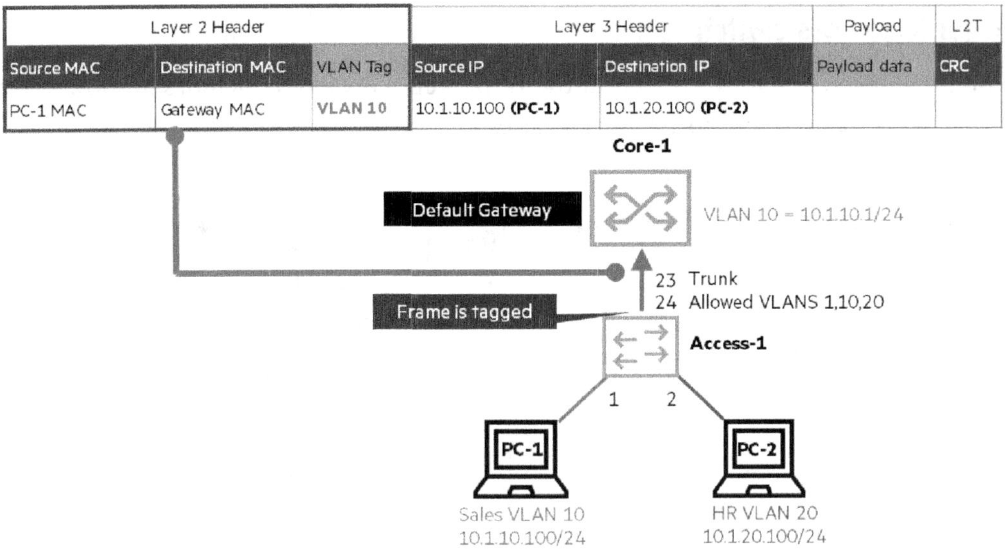

Figure 7-12: Access to multilayer switch

Routing process

Multilayer switch Core-1 is the Layer 2 destination of this frame. It accepts the frame, strips off the Layer 2 header, and begins to perform its routing function, to analyze Layer 3 header information.

It compares the Layer 3 destination IP address to its routing table entries. The figure shows Core-1's route table—the output of `show ip route`. Core-1 knows that destination network 10.1.20.0/24 is directly connected on its Switch Virtual Interface (SVI) VLAN20. Thus, Core-1 knows that it must forward the packet out its VLAN20 interface.

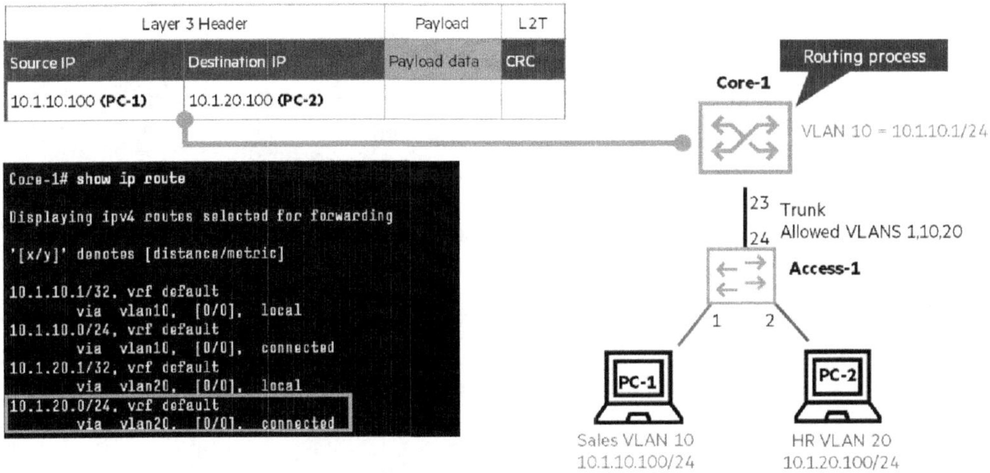

Figure 7-13: Routing process

Multilayer to access switch

Core-1 builds a new frame to wrap around the IP packet. This frame includes an 802.1Q tag—VLAN = 20. This frame is sent to L2 switch Access-1.

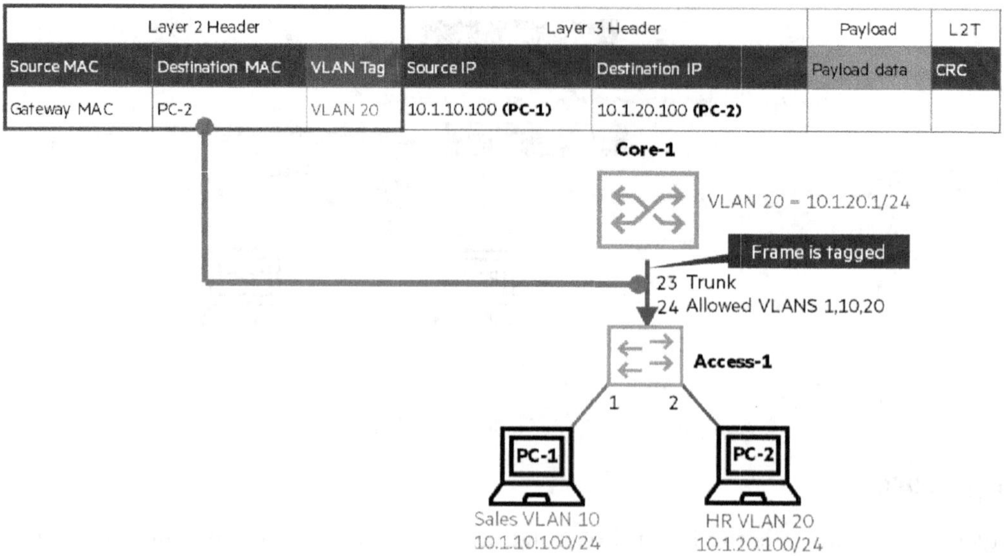

Figure 7-14: Multilayer to access switch

Access switch to endpoint

Access-1 receives the frame and learns from the tag that it is for VLAN 20. The 802.1Q tag has served its purpose, and so it is removed. Access-1 compares the destination MAC address to its MAC address table, finds a match, and so forwards the frame out port 2, toward PC-2.

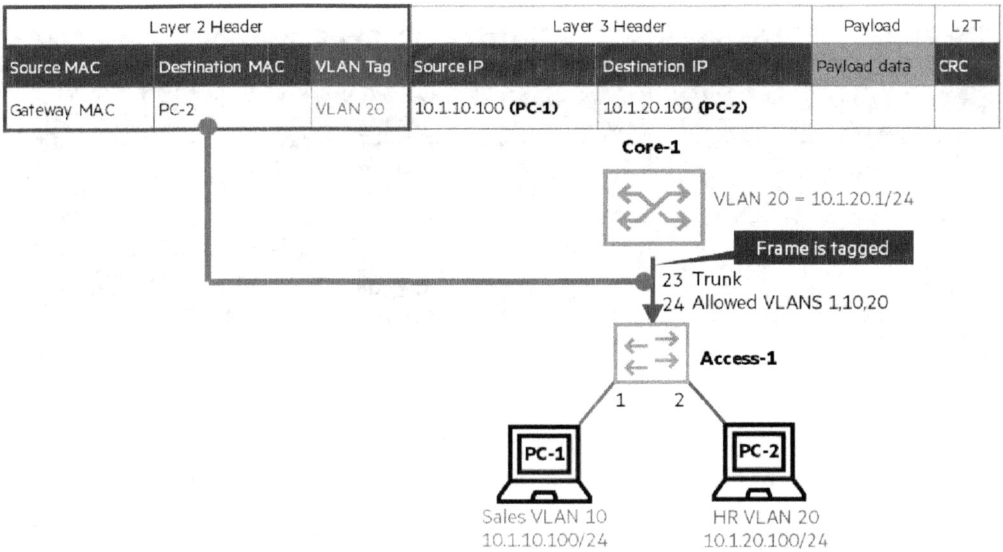

Figure 7-15: Access switch to endpoint

Virtual Routing and Forwarding

In Layer 3 routing, Virtual Routing and Forwarding (VRF) is a technology that enables the creation of multiple virtual routing instances within a single physical router. Each VRF instance operates as an independent routing domain, allowing the router to maintain separate tables and forwarding decisions for different network segments or customers.

VRF purpose

VRF enables the switch to logically segment its routing operations, effectively creating multiple virtual routers within a single physical device. Each VRF instance maintains its own routing table, forwarding decisions, and routing protocols, allowing for isolated routing environments that can serve different network segments, tenants, or applications without interface. This feature enhances network scalability, security, and resource utilization, particularly in multitenancy or service provider environments.

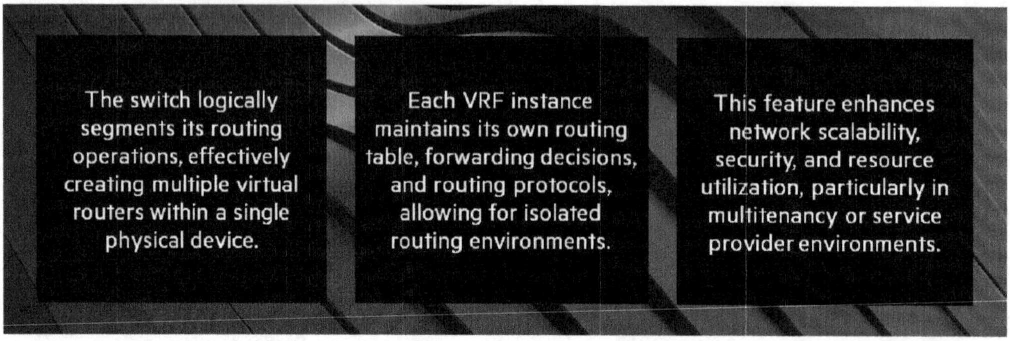

Figure 7-16: VRF purpose

You learned that you can define several VLANs on a single physical switch. It is as if you have created multiple virtual switches inside the physical switch—one for each VLAN.

Similarly, you can create separate virtual routers inside a single physical router, with VRF.

VRFs are useful in situations where the IP addressing overlaps in different places of the network. This could happen when two companies merge, for example. Also, another great use of VRF in a multitenant environment: customers A, B, and C could be physically connected to the same Layer 3 switch, but their routing tables are virtually kept separate and without using any ACLs to prevent communication between the customers. AOS-CX use **default** VRF for global routing and **mgmt** VRF for OOBM.

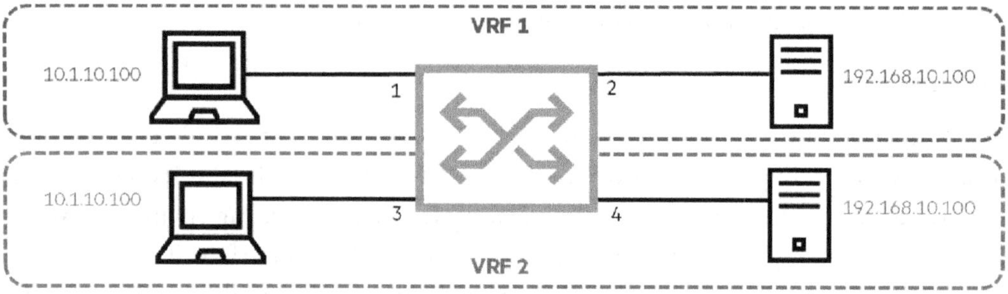

Figure 7-17: A single multilayer switch split into two separate VRFs

Interfaces 1 and 2 participate in VRF 1, and only interfaces 3 and 4 participate in VRF 2. These two VRFs do not interact. It is as if they are separate physical routers, with no connectivity between them. This is why the addressing can be the same in both VRFs without conflict.

In AOS-CX, all interfaces (enabled with routing) by default are mapped to the Global VRF called default. In other words, all interfaces are part of the same VRF—the physical router and the global VRF are essentially the same thing. Then you decide to create VRF 1 and 2, to split it up as shown in the figure. You know the two VRFs do not interact by default. However, you can configure the solution to route between the two VRFs if needed.

AOS-CX also includes a specific VRF for management purposes, and that can only be used in the Out-of-Band Management (OOBM) port to separate the data and control plane from the management plane.

Learning check

Check your understanding of this material by answering the following questions.

Question 1

A router's IP routing table has an entry with a next-hop IP address of 10.30.233.1. What does this number represent?

- ○ The packet destination.
- ○ The next Layer 2 switch that should receive the packet.
- ○ The cached ARP entry for the destination MAC address.
- ○ The next Layer 3 device that should receive the packet.
- ○ The router's own egress interface for routing.

Question 2

Which are true for providing dynamic IP addresses (DHCP) for hosts in a VLAN? (Select two)

- ☐ Using the IP helper feature in a multilayer switch.
- ☐ AOS-CX switches do not support hosts with dynamic IP addresses.
- ☐ A DHCP server could be attached (connected) to the same VLAN.
- ☐ Using static routes, a AOS-CX switch can forward DHCP discovery packets to the DHCP server.

8 IP routing fundamentals

Welcome to the IP routing fundamentals chapter. In this chapter, we are going to explore the foundational concepts of routing in computer networks. We will look at the fundamental principles of IP routing, including routing tables, routing protocols, and the role of routers in directing data packets across the networks. By understanding the essential components and processes of IP routing, you will gain the knowledge and skills needed to design, implement, and troubleshoot efficient routing infrastructures.

OBJECTIVES

After completing this chapter, you should be able to:

- ✓ Identify the fundamental principles underlying routing.
- ✓ Describe the structure and functions of routing tables.
- ✓ Differentiate between Distance Vector and Link State routing protocols.
- ✓ Configure static and dynamic routing protocols on network devices.

Fundamental principles of routing

To begin, we are going to explore the various types of routes, from directly connected routes to remote routes, and their significance in routing decisions. Additionally, we will examine the concept of administrative distance and its role in determining the priority of routes in routing tables.

Routing, in computer networks, refers to the process of directing data packets from their source to their destination across interconnected networks. It involves determining the optimal path for data transmission based on network topology, addressing information, and routing protocols. The goal of routing is to ensure efficient and reliable communication between devices within a network and across different networks on the internet.

Importance of routing

Understanding the importance of routing is crucial for network administrators and engineers as it enables you to design, implement, and troubleshoot efficient and reliable network infrastructures, ensuring optimal data transmission and connectivity between devices and across interconnected networks. Additionally, knowledge of routing principles allows organizations to make informed decisions about network architecture, routing protocols, and traffic management strategies, ultimately contributing to the overall performance and security of your networks.

There are several factors that make understanding routing important.

- **Data transmission:** Routing determines the path that data packets take from their source to their destination across interconnected networks. By directing packets along the most efficient routes, routing ensures timely and reliable delivery of data between devices.
- **Network connectivity:** Interconnected networks, such as the internet, consist of numerous routers and network segments. Routing protocols enable these devices to exchange routing information and dynamically adapt to changes in network topology, ensuring continuous connectivity and resilience against network failure.
- **Scalability:** As networks grow in size and complexity, routing enables scalable communication by efficiently handling traffic across multiple network segments and routing paths. This scalability allows networks to accommodate increasing numbers of devices and users without sacrificing performance or reliability.
- **Fault tolerance:** Routing protocols support redundant network paths and alternate routes, allowing data packets to be rerouted in the event of link failures. This fault tolerance ensures that communication remains uninterrupted even when the individual network components experience issues.
- **Traffic engineering:** Routing enables network administrators to optimize traffic flow and resource utilization by implementing traffic engineering policies. By configuring routing protocols and route preference, administrators can prioritize certain types of traffic, balance network loads, and minimize latency for critical applications and services.

In summary, routing is essential for establishing and maintaining communication between devices in interconnected networks.

Route types

Understanding route types is fundamental to effective network design and management. A routing device makes Layer 3 forwarding decisions based on the best-path entries in the routing table. The entries in the routing table are populated manually or through a routing protocol.

There are multiple types of routes in computer networking; for the purposes of this chapter, we will focus on AOS-CX route types.

Route types that can exist in an AOS-CX switch include the following:

- **Directly connected routes:** Routes to the networks that are directly connected to the router's interfaces. The subnet is physically connected to the device.
- **Local:** The subnet is configured inside the device (loopback or SVI).
- **Static routes:** Routes that are manually configured by the network administrator and do not change unless modified. (Default and Floating)
- **Dynamic:**
    - RIPv2: route learned via Routing Information Protocol v2
    - OSPF: route learned via Open Shortest Path First
    - BGP: route learned via Border Gateway Protocol

Connected and local routes

Connected routes represent networks directly connected to the router's interfaces, while local routes refer to specific host IP addresses within the local network. Both types of routes enable communication between devices within the network.

Consider the following configuration, where the Switching Virtual Interface (SVI) for VLAN1 receives an IP address and port 1/1/1 is assigned to VLAN 1.

```
Switch(config)# interface vlan 1
Switch(config-if)# ip address 10.1.1.1/24
Switch(config)# interface 1/1/1
Switch(config-if)# routing
Switch(config-if)# vlan access 1
```

This configuration generates some entries in the routing table; you can verify it using the `show ip route` command. In this case, you can see that two entries are created in the routing table.

The first entry is listed as 10.1.1.0/24 and is a connected type route. This entry means that the subnet is available physically connected to the switch and there is no need for a next-hop device.

The second entry is listed as 10.1.1.1/32 and is a local type route. This entry means that the IP address that was previously configured is available for the routing process inside the switch. This could be a loopback interface, or as shown here, a Switch Virtual Interface (SVI) for VLAN 1.

You may notice that the local entry uses a /32 subnet mask. This is a special mask that indicates that subnetting cannot happen (no divisions are permitted) in the address space provided. This mask is used to refer to an individual host address. In this case, only one IP address is permitted to be configured in the SVI.

Static routes

Static routes are manually configured routes in a router's routing table that do not change unless modified by the network administrator. They specify the path to reach specific destination networks or IP addresses and are entered into the router's configuration either manually or through a script.

Figure 8-1: Key characteristics of routes

Key characteristics of static routes include:

- **Static configuration:** Network administrators manually configure static routes on router-s/switch routers specifying the destination network or IP address, as well as the next-hop router or exit interface to reach that destination.
- **Predictable routing behavior:** Static routes remain unchanged unless explicitly modified by the administrator. This predictability makes them suitable for networks with stable topologies or specific routing requirements.
- **No dynamic updates:** Unliked dynamic routes, static routes do not automatically update in response to network changes or updates. They require manual intervention to modify or remove, making them less adaptable to changes in network conditions.

- **Lower overhead:** Static routes typically have lower processing overhead compared to dynamic routing protocols, as they do not involve periodic routing updates or dynamic route calculations. However, since static routes must be routed manually, they have a heavy load for network administrators.
- **Use cases:** Static routes are commonly used to define specific paths for traffic destined to or from particular networks, to configure default routes for outbound traffic, or to create backup routes for redundancy.

In AOS-CX, use the `ip route` command to build a static route.

```
Switch(config)# ip route <REMOTE-NETWORK/MASK> <NEXT-HOP> vrf <VRF-ID>
```

Example

Suppose that you need endpoints in subnet 10.1.10.0 to communicate with others in 10.1.20.0. The Core-1 switch has two physically connected networks: 10.1.10.0/24 and 10.1.12.0/30. However, the switch is not physically connected to the 10.1.20.0/24 subnet. So, you add the following route:

```
Core-1(config)# ip route 10.1.20.0/24 10.1.12.2
```

The Core-2 switch also needs a static route to be able to reply to any endpoints on network 10.1.10.0/24. From its perspective, it only has entries in its routing table for networks 10.1.20.0/24 and 10.1.12.0/30. You must add a new route:

```
Core-2(config)# ip route 10.1.10.0/24 10.1.12.1
```

To verify this entry exists on the routing table, issue the `show ip route` command.

Floating static routes

Floating static routes are backup routes configured with a higher administrative distance than primary routes. These routes are designed to become active only if the primary routes fail or become unavailable. Manipulating the administrative distance can be used to create primary and backup routes, to improve network resiliency.

Key points about floating static routes include:

- **Higher administrative distance:** Floating static routes have a higher administrative distance compared to primary static routes. This higher value makes them less preferred by routers under normal conditions.
- **Triggered activation:** Floating static routes become active and are used for routing traffic only when the primary routes fail. This failure could be due to link outages, network congestion, or other network issues.

- **Redundancy and failover:** Floating static routes provide redundancy and failover capabilities in network environments where continuous connectivity is critical. By having backup routes ready to activate in case of primary route failures, organizationscan ensure uninterrupted network communication and minimize downtime.
- **Manual configuration:** Like standard static routes, floating static routes are manually configured by network administrators. They are typically configured with higher administrative distances to distinguish them from primary routes.
- **Use cases:** Floating static routes are commonly used in scenarios where redundancy and failover are essential, such as in high-availability networks, critical infrastructure deployments, or in conjunction with dynamic routing protocols to provide backup paths.

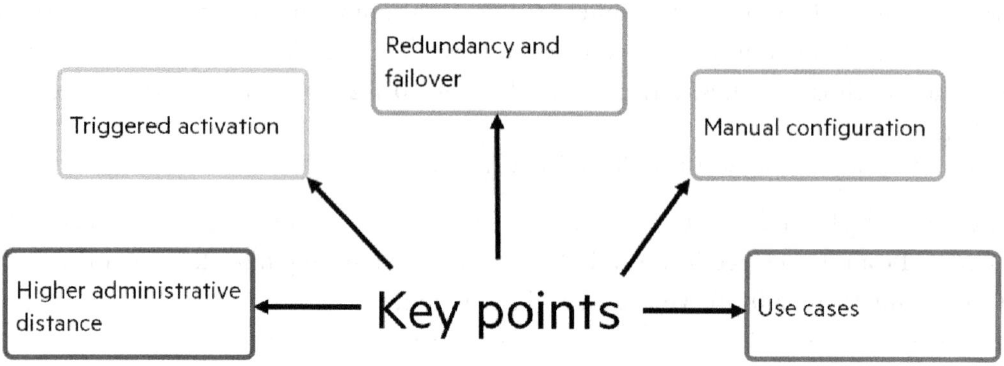

Figure 8-2: Floating static route key points

Example

Imagine that you have two internet connections that use different ISPs. The first connection has a higher bandwidth than the second one, and therefore, the administrator wants to use it as the primary connection. The secondary ISP connection will be a backup, waiting to be used in case the primary link fails.

To do this, configure two static routes, one to the "all routes" destination of 0.0.0.0 with ISP-1 as the next-hop. You let its default administrative distance remain at 0.

Configure a second static route to the "all routes" destination of 0.0.0.0 with ISP-2 as the next-hop, changing the administrative distance to 10.

```
Core-1[config]# ip route 0.0.0.0/0 10.1.251.2
Core-1[config]# ip route 0.0.0.0/0 10.1.252.2 distance 10
```

Since the ISP-1 route has a lower administrative distance, it will be used exclusively. The ISP-2 path shall only be used if the primary link fails. You can verify this behavior using the `show ip route` command.

Network 0.0.0.0 with a subnet mask of 0.0.0.0 is a super network that summarizes all possible addresses in the IPv4 scope—meaning any address.

It is important to differentiate between default gateway and default route. The first concept refers to a device that does not have the routing feature enabled, such as a host or a Layer 2 switch. In that situation, we just provide the information about where to send packets that are not in the same subnet. A default route defined as 0.0.0.0/0 is used in devices where the routing process is enabled, such as in routers and multilayer switches.

Introduction to dynamic routes

Dynamic routes, integral to modern networking, are routes automatically learned and updated by routers through dynamic routing protocols, facilitating efficient and adaptive routing decisions based on real-time connections.

Dynamic routes have several key characteristics that distinguish them from static routes:

- **Automated updates:** Dynamic routes are automatically learned and updated by routers through dynamic routing protocols. This automation eliminates the need for manual configuration and interventions by network administrators, saving time and effort in managing routing information.
- **Adaptability to network changes:** Dynamic routes are responsive to changes in network topology, link status, and routing metrics. Routers exchange routing updates with neighboring routers, allowing them to dynamically adjust their routing tables to reflect changes in the network, such as link failures or new network connections.
- **Scalability:** Dynamic routing protocols are designed to scale efficiently to large and complex networks. They can handle networks with hundreds of thousands of routers and adapt to changes in network size and structure without significant performance degradation.
- **Route convergence:** Dynamic routing protocols facilitate route convergence, ensuring that all routers in the network converge on consistent routing information within a relatively short time frame. This rapid convergence reduces the likelihood of routing loops or suboptimal routing paths in the network.
- **Routing metrics:** Dynamic routing protocols use various metrics, such as hop count, bandwidth, delay, or path cost, to determine the best path to reach a destination network. These metrics help routers select the most efficient routes based on network conditions and performance criteria.

- **Load balancing:** Dynamic routing protocols support load balancing by distributing traffic across multiple paths to reach a destination. This helps optimize network utilization and improve overall network performance by preventing congestion on individual links.
- **Protocol specific characteristics:** Different dynamic routing protocols have unique characteristics and features tailored to specific network environments and requirements. For example OSPF, BGP, and RIPv4.

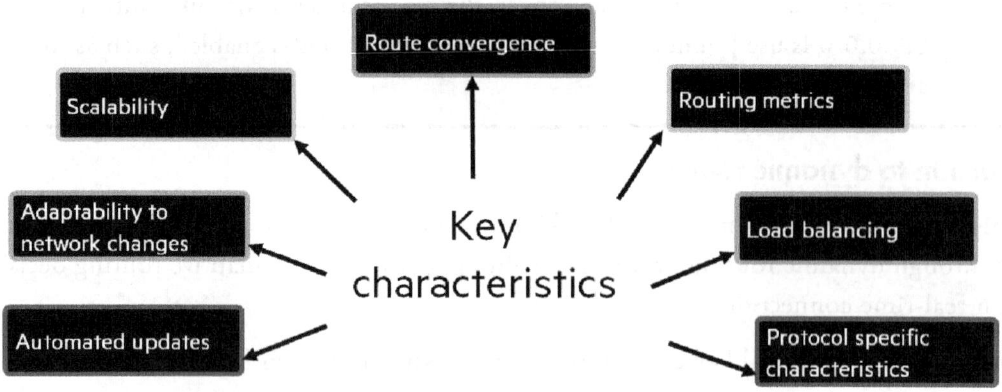

Figure 8-3: Key characteristic of dynamic routes

Static versus dynamic comparison

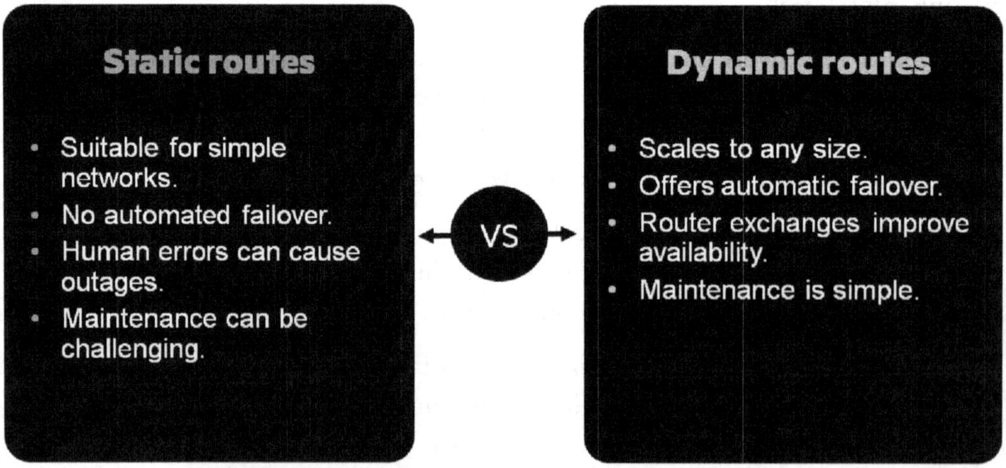

Figure 8-4: Comparison of static versus dynamic routes

Working with static routes is simple and easy for small networks that have a few subnets. However, when an organization has hundreds or even thousands of subnets, static routing is not an efficient method to administer subnets. There is no automatic route advertisement; you must manually configure everything. Other than the simple floating static routes you just learned, there is no dynamic failover mechanism.

The human factor is also a big consideration. Administrators can cause serious problems if a route is placed in the incorrect device or if the route is not properly configured. This can create Layer 3 route loops, traffic blackholing, and lost connectivity.

Recovering from a routing device failure could be slow if static routing is used. You must manually configure new alternative paths. This is inconvenient at best, especially when dynamic routing would automatically do this for you.

Dynamic routing protocols are far more scalable, handling thousands or even millions of routes across multiple routing devices. They automatically fail over to alternate paths with little to no downtime. This is because routers constantly exchange messages that keep the network available and minimize your managerial workload.

Routing tables

While we covered routing tables in the previous chapter, it never hurts to do a review and emphasize the structure and routing.

Remember, a routing table is a critical component of a router's functionality, serving as a map that guides the router in forwarding data packets to their intended destination across interconnected networks. Here is a review of routing tables, including their structure and function.

Routing tables consist of entries, each representing a route to a specific destination network or IP address. These entries typically include:

- **Destination network:** The network address or IP prefix of the destination network.
- **Subnet mask:** The subnet mask associated with the destination network, indicating the network's address range.
- **Next-hop:** The IP address of the next-hop router or exit interface to which packets should be forwarded to reach the destination network.
- **Interface:** The interface through which packets should be forwarded to reach the next-hop router or directly connected network.
- **Administrative distance:** A numerical value assigned to the route, indicating its trustworthiness or priority relative to other routes.
- **Cost:** A numerical value assigned to the route, indicating the priority of a route relative to other routes from the same routing protocol. The cost can be used as a tie-breaker when two or more routes are available to the same destination using the same routing protocol and same administrative distance.

Administrative Distance (AD) is a metric used in routing to determine the trustworthiness or reliability of routing information from various sources. It is a numerical value assigned to routing information by a router to indicate the "trustworthiness" of a route learned from a particular source. The lower the administrative distance, the more reliable or preferred the route is considered by the router.

In practical terms, administrative distance helps routers determine which route to select when multiple routing protocols or sources provide information about the same destination network. If a router receives routing updates from multiple sources (for example static routes, dynamic routing protocols), it uses administrative distance to prioritize the routes in its routing table. The router selects the route with the lowest administrative distance as the more preferred route to reach the destination network.

Each routing source or method has a predefined administrative distance assigned to it.

Directly connected routes typically have an administrative distance of 0, indicating the highest trustworthiness.

Static routes often have an administrative distance of 1 or a lower value, making them more preferred than routes learned from dynamic routing protocols.

Routes learned from dynamic routing protocols like OSPF or BGP have administrative distances assigned based on their protocol-specific characteristics. These values are typically higher than those for directly connected or static routes.

Sometimes a router will learn a route from two different sources. Maybe a router is running BGP and OSPF, and you have entered some static routes. Perhaps all three methods have taught the router about network 10.1.20.0/24. Which source for routing information should the router trust? The source with the lowest administrative distance.

Table 8-1: Administrative distances for each routing protocol

Populating method	Administrative distance
Directly connected	0
Static route	1
External BGP	20
OSPF	110
Internal BGP	200
Unknown or unused	225

Routing protocols

In this section, we explore the diverse landscape of routing protocols, each designed to address specific network requirements and challenges. From interior gateway protocols like OSPF to exterior gateway protocols such as BGP, routing protocols play a vital role in shaping the dynamic behavior of modern networks.

Interior and exterior gateway protocols

An Autonomous System (AS) is a collection of routers under a common administrative domain. Your Internet Service Provider (ISP) owns their own internal network and it is their AS. Each corporation has autonomy over their network. It is their AS.

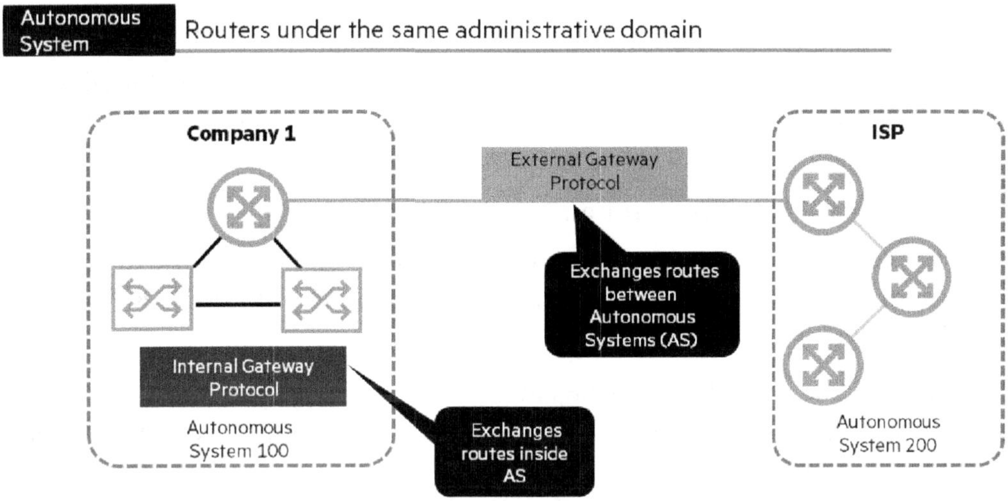

Figure 8-5: AS using RIP to exchange network information and find the best path

To route packets inside an AS, each company chooses an Interior Gateway Protocol (IGP). An IGP is simply a routing protocol that runs inside an AS. Examples of IGPs include Routing Information Protocol (RIP), Intermediate System to Intermediate System (IS-IS), and Open Shortest Path First (OSPF). Each company can use the IGP that works best for them.

In the previous figure, AS 100 has three routing devices. They use RIP to exchange network information to discover all paths inside the AS and then find the best path for each network.

Distance vector routing protocols

The earliest methods devised to exchange route information used Distance Vector algorithms. Each router advertises its distance from each network, and the vector or direction packets must travel to get to that network.

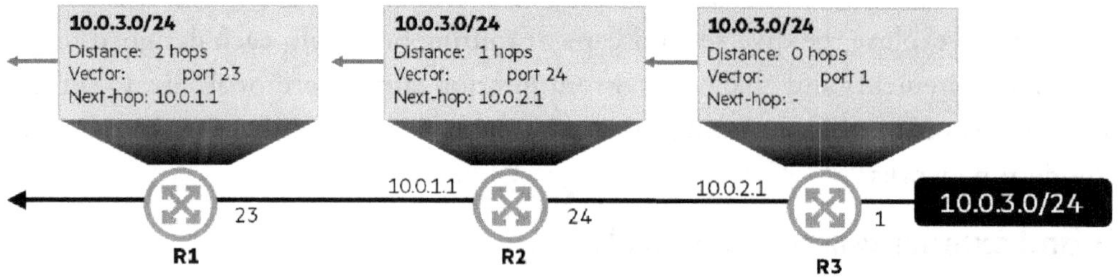

Figure 8-6: Distance vector routing with next-hops

The distance is measured in "hop count." In the figure, how many routers must R3 "hop over" to get to network 10.0.3.0/24? The answer is no routers—it is directly connected to that network.

What vector or direction must R3 forward packets to reach 10.0.3.0/24? The answer is "out port 1." It is as if the router says, "To get to any host on destination network 10.0.3.0/24, which is zero hops away, I must forward packets out my local port 1." Using something like RIP or RIPv2, R3 advertises its distance and vector to R2.

R2 receives this and determines that "If R3 is zero hops away, and I must hop over R3 to get to that network, then I am one hop away." To get to destination 10.0.3.0/24, which is one hop away, R2 must forward packets out local port 24, to next-hop router 10.0.2.1. R2 advertises its distance and vector to the next router, R1.

R1 receives this and knows, "If R2 is one hop away, and I must hop over R2 to get to the destination, then I am two hops away." To get to 10.0.3.0/24, R1 forwards packets out local port 23, to next-hop router 10.0.1.1.

Distance Vector routers are not aware of the entire network topology. They only perceive their directly connected routing peers. This simplifies operation, but also limits operation:

- **Slow convergence:** When a failure occurs, Distance Vector protocols can take minutes to converge, depending on network size, complexity, and architecture.
- **Limited scalability:** Each router can be no more than 15 hops away from any other router. Understand that you could still have dozens or even hundreds of routers, as long as it is a relatively flat architecture.

Distance Vector protocols include the Routing Information Protocol (RIP), RIP version 2 (RIPv2), and RIP Next Generation (RIPng). RIP and RIPv2 do not support security features to protect the advertisement messages that are exchanged between routing devices.

Due to their limited scalability, performance, and security, AOS-CX does not support any Distance Vector protocols. Instead, AOS-CX supports the more robust Link State routing protocols.

Link-state routing protocol

Link-State protocols resolve limitations and disadvantages of the Distance Vector protocols. Routers that run Link-State protocols possess information about the complete network topology. This approach enables them to independently calculate their routing table using the more sophisticated Dijkstra algorithm to select the best path for every network destination. The Dijkstra algorithm has fast convergence and is scalable.

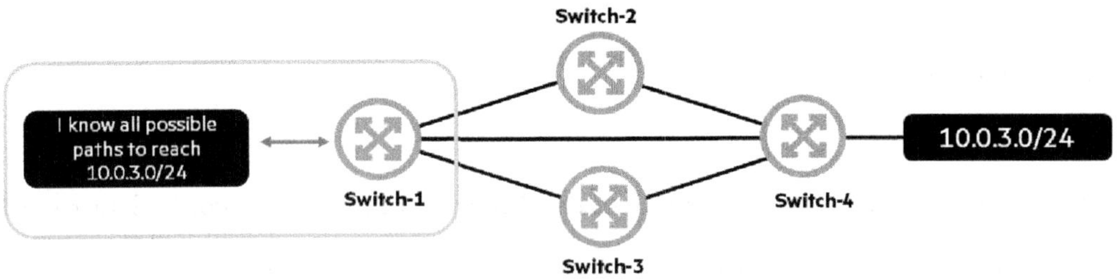

Figure 8-7: Link-state routing protocol using OSPF for routing

Actual implementations of Link-State protocols include Open Shortest Path First version 2 (OSPFv2), which is used for IPv4; OSPFv3, which is used for IPv6, and Intermediate System to Intermediate System (IS-IS). IS-IS is far less common than OSPFv2, and so this chapter is focused on OSPF. These protocols provide faster convergence times compared with Distance Vector protocols. Another advantage is scalability.

One disadvantage of protocols like OSPF is that they create more CPU, memory, and bandwidth utilization than protocols like RIP. Many years ago, this was an issue; however, routers have become more capable over the years, and our ability to design more efficient network architectures has greatly reduced this concern. Furthermore, the AOS-CX multilayer switches are designed to meet today's network requirements. You can verify this information by running the `show capacities ospfv2` command.

```
Switch-6300# show capacities ospfv2
System Capacities: Filter OSPFv2
Capacities Name                                                      Value
--------------------------------------------------------------------------
Maximum number of OSPFv2 areas configurable in the system              128
Maximum number of OSPFv2 interfaces configurable in the system         128
Maximum number of OSPFv2 interfaces per area in the system             128
Maximum number of OSPFv2 neighbors allowed across all VRFs             128
Maximum number of OSPFv2 processes supported across each VRF             1
Maximum number of routes in OSPFv2 supported across all VRFs         64000
```

Learning check

Check your understanding of this material by answering the following questions.

Question 1

Suppose that a router has learned about network 172.18.37.0/24 from three sources – OSPF, Internal BGP, and a static route. Which statement is true about this router's path selection?

- O The router chooses the OSPF route, because it is trusted more than BGP.
- O The router chooses the best path based on the lowest cost.
- O The router chooses the option with the highest administrative distance.
- O The router chooses the static route.

Question 2

Which of the statements below accurately describe link state routing protocols? (Select two)

- ☐ They can be used to route packets between Autonomous Systems.
- ☐ They can be used to route packets inside an Autonomous System.
- ☐ They use a Distance Vector algorithm.
- ☐ They are more scalable than protocols like RIP.
- ☐ OSPFv2 is a very popular choice for Exterior Gateway Protocol-based routing.

Question 3

What is the administrative distance of a directly connected route?

- O 0
- O 1
- O 90
- O 100

9 Dynamic IP Routing - Open

Welcome to the Dynamic IP Routing - Open Shortest Path First (OSPF) chapter. OSPF is a routing protocol that operates at the network layer (Layer 3) and is used to exchange routing information between routers in IP networks. OSPFv2 is a specific version that supports IPv4 addressing and is the most commonly deployed. It calculates the shortest path to each destination network based on IPv4 addresses and exchanges routing information using IPv4 headers.

OSPF single area, also known as single-area routing, refers to a network configuration where all OSPF routers and networks reside within a single OSPF area. In OSPF, an area is a logical grouping of routers and networks that share common routing information and policies.

This chapter is going to give you an overview, or foundational learning, of OSPF. In the hands-on lab activity, you will learn how to configure OSPF. In the *Implementing AOS-CX Switching* and *Advanced AOS-CX Switching* guides, the information and lab exercises will build upon the knowledge you learn in this chapter.

OBJECTIVES

After completing this chapter, you should be able to:

- ✓ Describe OSPF.
- ✓ List the various phases in OSPF general operation.
- ✓ Explain the various steps involved in neighboring.
- ✓ Identify the various types of OSPF networks.
- ✓ Explain the need for selecting Designated Router (DR), Backup Designated Router (BDR), and DROTHER.
- ✓ Explain OSPF LSA types, path selection, and convergence.
- ✓ Configure OSPF in AOS-CX.

Introduction to OSPFv2

Open Shortest Path First (OSPF) is a protocol that was developed to route Internet Protocol (IP) packets within a single routing domain. In this chapter, you will learn the basics of OSPFv2, which is designed for IPv4 networks. Request for Comments (RFC) is a technical publication from the Internet Engineering Task Force (IETF). RFC 2328 defines OSPFv2 to calculate routes for IPv4 packets. OSPF does not use TCP or UDP—OSPF advertisements are placed directly inside an IP packet. Therefore, it does not have a TCP or UDP port number; it has IP protocol number 89.

The IP protocol number is a number associated to the protocol that is in use in Layer 4. The IP protocol number is included as a field in the Layer 3 header. This announcement in Layer 3 helps network devices to be aware of the Layer 4 protocol that is in use without decapsulating the packet. For comparison, TCP uses protocol number 6 and UDP uses protocol number 17.

OSPF is a very popular enterprise Interior Gateway Protocols (IGP) routing solution due to its hierarchical scalability and security mechanisms. For example, OSPF peers can authenticate packet exchanges. As a link-state protocol, OSPF-enabled routers advertise information about their connected Layer 3 interfaces and networks (links) and the cost associated with those interfaces. OSPF collects link state information from available routers to construct a topology map of the network. OSPF divides the network into routing areas, which are identified by 32-bit numbers.

The primary characteristics of OSPF are:

- **Dynamic routing:** OSPF enables routers to dynamically exchange information about network routes, allowing them to build and maintain accurate routing tables based on the current network topology. This dynamic routing capability ensures efficient and adaptive forwarding of IP packets across the network.
- **Scalability:** OSPF is well suited for larger and complex networks, as it supports hierarchical routing and can scale to accommodate networks with thousands of routes and subnets. It divides networks into areas to reduce the size of routing tables and minimize routing traffic, improving scalability and network performance.
- **Fast convergence:** OSPF uses link-state routing algorithms to calculate the shortest path to each destination network, leading to fast convergence and optimal path selection. It quickly adapts to changes in the network topology, such as link failures or additions, ensuring rapid re-routing and minimal disruption to network traffic.

- **Load balancing:** OSPF supports Equal Cost Multipath (ECMP) routing, allowing routers to distribute traffic evenly across multiple paths with the same cost. This load balancing capability helps optimize network utilization and improve overall network performance.
- **Security:** OSPF provides mechanisms for authenticating routing updates and protecting against unauthorized access and malicious attacks. Authentication ensures the integrity and authenticity of routing information exchanged between OSPF routers, enhancing network security.

OSPF plays a critical role in facilitating efficient and reliable IP routing in large-scale networks, providing scalability, fast convergence, load balancing, and security features to support the dynamic nature of modern networks.

Router ID

All OSPF routers require a unique Router Identifier (RID), which is a 32-bit dotted-decimal notation address. Routers include their ID in all OSPF packets that they send.

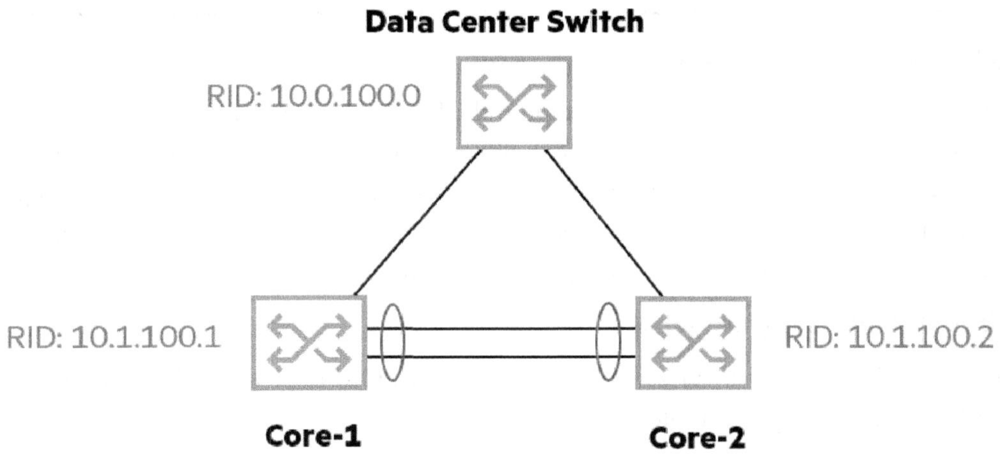

Figure 9-1: A data center switch and two other switch interconnected

You might look at this topology and see "Core-1, Core-2, and Data Center Switch." The routers might identify themselves as 10.1.100.1, 10.1.100.2, and 10.0.100.0.

You can depend on each router to automatically identify itself, or you can take control of the situation and manually assign RIDs. Automatic assignment is tempting, since it requires little effort. However, many experienced engineers prefer to manually assign RIDs due to certain documentation, troubleshooting, and management advantages. Either method is acceptable.

AOS-CX uses the following sequence to determine the router ID:

1. If you manually specify the RID, then that is what the router uses.
2. If you do not specify the RID, the loopback interface with the highest IP address is assigned as the RID.
3. If no loopback interfaces exist, the regular non-loopback interface with the highest IP address is assigned as the RID. Non-functional interfaces that are in a down state are not considered.

A loopback interface is defined as a logical interface that is always in an up state (unless you manually disable it). This interface is useful for processes and protocols that depend on the interface status to work. The IP address associated to a loopback interface is routable, which means that external devices can initiate communication to it. As you gain more education and experience, you will continue to learn about the advantages of loopback interfaces for scalability, troubleshooting, and network management.

OSPF general operation overview

A router's job is to learn about every path, choose the best paths, and then deploy those paths to the routing table. To do this, OSPF uses a four-phased approach. Each phase results in a kind of table that is built, maintained, and stored in local memory on each router. The more deeply you understand each phase and its associated table, the better you will be at designing, deploying, and troubleshooting OSPF networks. The four phases are as follows:

- **Phase 1:** Build a neighbor table

  Each OSPF router establishes a relationship with an OSPF neighbor. This information is available in the neighbor table.

- **Phase 2:** Build a topology database

  The topology database is also called a Link State Database (LSDB). OSPF-enabled routers exchange network prefixes and parameters. This information is stored in the LSDB table.

- **Phase 3:** Build an OSPF route table

  Each OSPF-enabled router runs the Shortest Path First (SPF) algorithm to get the best path for each network destination. The result populates the OSPF routing table.

- **Phase 4:** Build a routing table

  In this phase, the OSPF internal process publishes its results into the routing table (which is also know as a Forwarding Information Base (FIB) that is used to forward packets.

We will dive deep into each of these four phases in the next section.

Phase 1 - Neighboring

Objective: Form a relationship with OSPF peers

Method: Send hello packets and validate compatibility

In this phase, first the directly connected OSPF neighbors introduce themselves to their directly connected neighbors. The objective is to validate that they are compatible. This is achieved by sending OSPF "hello" packets out each OSPF interface.

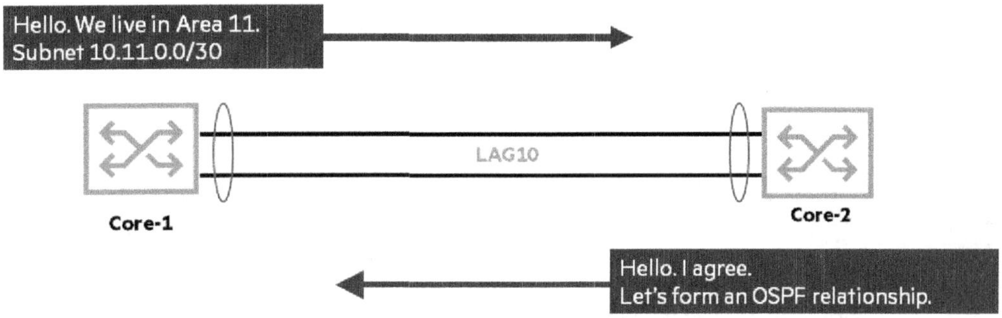

Figure 9-2: Two switches sending hello packets to form OSPF relationship

For example, Core-1 sends hellos out on all its interfaces where OSPF has been enabled. Core-1 sends a hello out on LAG10, on subnet 10.11.0.0/30; "Hello. I am Core-1. I think that we live in Area 11, a normal area, and we do not need to use secure authentication. We are connected on subnet 10.11.0.0/30." Core-2 sends similar information in its hello packet out on LAG10. As long as the criteria match, the OSPF routers agree to be neighbors, that is, they form an OSPF neighbor relationship. If any of these parameters do not match (typically due to mis-configuration), the routers refuse to form a neighbor relationship, and your network is broken. Let us assume that all routers in this example were successful in forming an adjacency with each directly connected peer. This will be reflected in each router's neighbor table.

Only a small part of the hello packet contents are shown and discussed here, to convey the main idea. There is much more information exchanged in hello packets, which you will soon learn.

The term "adjacency" refers to a couple of routers that are physically connected to each other. This term is different from OSPF neighboring because an OSPF neighbor relationship can be established between a couple of routers that are not physically connected.

Phase 2 - Build a topology database

Objective: Learn every link, router, and interconnection

Method: Send/receive multicast LSAs

A router shares known subnets with other routers if and only if an adjacency is fully formed. The objective is to build a topology database—a list of every single link (subnet), every single router, and exactly how those routers and subnets are interconnected. This is achieved by sending multicast Link State Advertisements (LSAs).

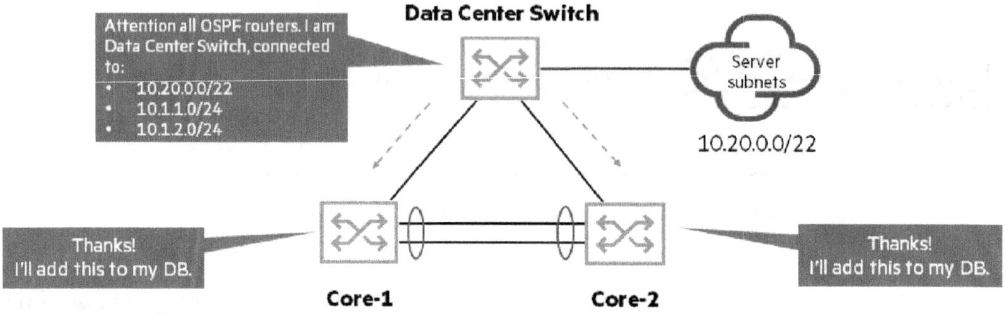

Figure 9-3: Switches sending out Link State Advertisements (LSAs)

In this example, Data Center Switch sends LSAs out on all of its interfaces: "Attention all OSPF routers. I am Data Center Switch, and I am directly connected to 10.20.0.0/22, 10.1.1.0/24, and 10.1.2.0/24." Every other OSPF router receives these LSAs and adds the information to their topological databases. Soon, every router has received LSAs from every other router. Thus, each router has a full topology database. Again, this database is essentially the diagram you see in the above example, in a numerical format. There is only one reality: the figure shown here in this example. Therefore, every OSPF router has an identical topological database.

The example only shows Data Center Switch advertising its directly connected subnets because that is all that it currently knows. However, once Data Center Switch receives LSAs from the other routers (as they have just received from Data Center Switch), it will advertise those routes as well. Routers advertise all the contents of their topology database to all other routers. The topology database is also known as the Link State Database (LSDB).

Phase 3 - Build OSPF route table

Objective: Build the best paths for each destination

Method: Run the SPF (Dijkstra's) algorithm on topology DB

Although there is only one topology, each router's position and connectivity within that topology are unique. Thus, each router's objective is to determine the best path to each link from where it sits within the topology. The method is to run the SPF (Dijkstra's) algorithm on the topology database or LSDB.

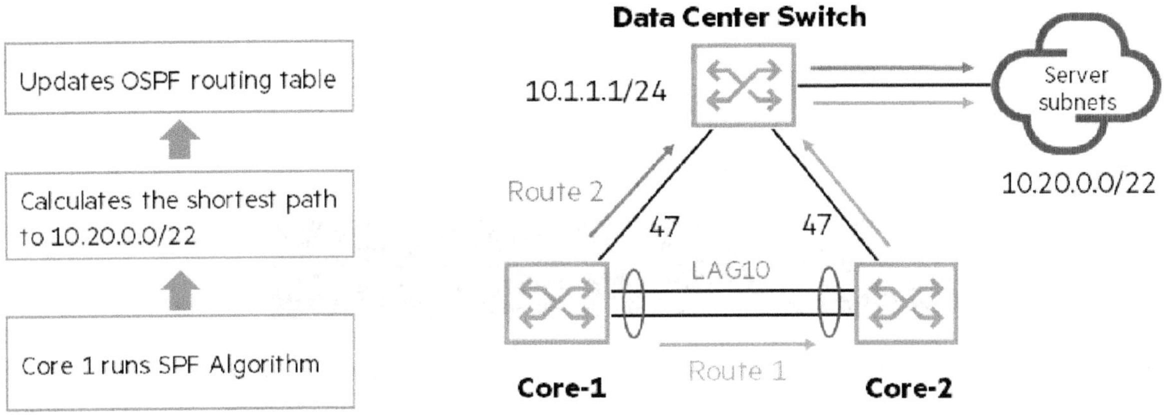

Figure 9-4: Calculating the shortest path first in a topology

For example, Core-1's SPF algorithm analyzes the topology database and sees that Core-2 could reach destination 10.20.0.0/22 by routing packets out port LAG10. Core 1 can also reach the same destination by routing packets out port 47. The algorithm considers the cost associated with each path (based on bandwidth) and chooses the lowest cost (typically fastest) path. This best path is added to the list of best paths: the route table. In this example, the best path is to send the traffic directly to Data Center Switch.

Phase 4 - Build a routing table

Objective: Publish the best path on the Forwarding Information Base (FIB)

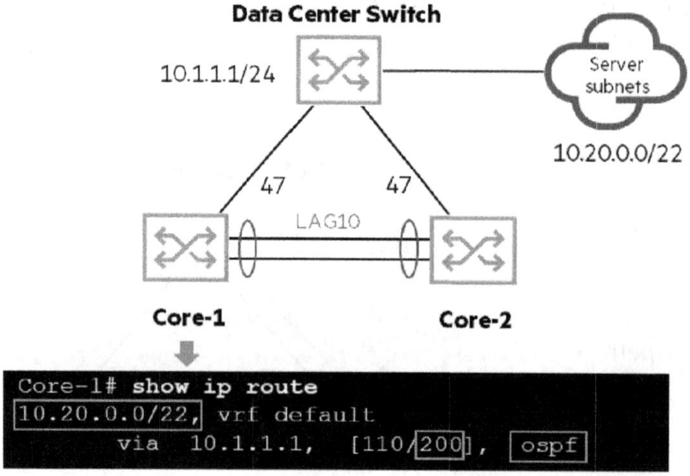

In this final step, the best path previously calculated by OSPF is published in the routing table (FIB). In this case, the best path from Core-1 to Server subnets uses the path through Data Center Switch.

Neighboring

In this section, we will learn about the hello messages that are sent in the neighboring phase and the various neighbor states.

Hello messages

Directly connected OSPF routers send hello packets to ensure two-way communication and to act as a failure detection mechanism. As in the following example, by default, these packets are sent every 10 seconds to Multicast IP address 224.0.0.5. This is the reserved "attention all OSPF routers" multicast address. Its associated MAC address is 01:00:5E:00:00:05.

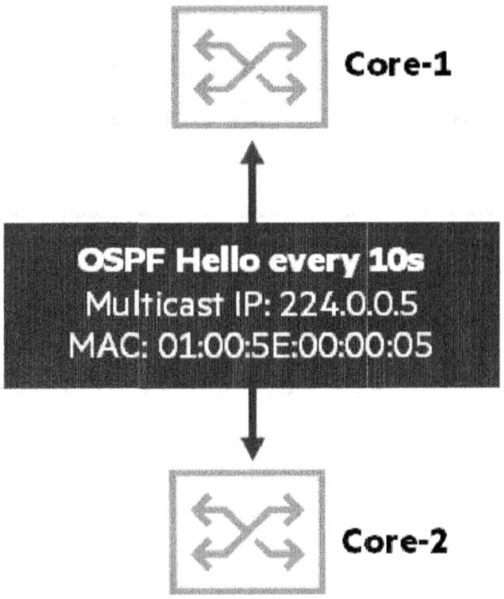

Figure 9-5: Hello message between two switches

Layer 2 switches forward broadcast and multicast frames out all ports in the broadcast domain, so all directly connected routers exchange hellos, the first packet to be sent in the OSPF exchange process. Another main purpose of hello packets is to build and maintain a neighbor table. Peers only form a neighbor relationship if they are compatibly configured. They must agree that they are on the same subnet and have the same subnet mask. They must be in the same area and agree on the area type. Their timers must match (10-second OSPF hello timer, etc.), and they must be configured to use the same authentication type.

To verify the hello interval, you can use the `show ip ospf neighbors detail` command.

```
Core-2# show ip ospf neighbors detail
Neighbor 10.11.100.1, interface address 10.11.0.1
--------------------------------------------------
Process ID 11 VRF TABLE-11, in area 0.0.0.11 via interface vlan110
Neighbor priority is 1, State is FULL
DR is 10.11.0.1, BDR is 10.11.0.2
Options is 0x42
Dead timer due in 00:00:30
Retransmission queue length 0
Time since last state change 00h:02m:39s
```

In AOS-CX, the hello interval is 10 seconds (by default). However, you can customize the interval from 1 to 65,535 seconds.

```
Switch(config)# interface <INTERFACE-ID>
Switch(config-if)# ip ospf hello-interval <SECONDS>
```

The dead interval is the interval of time after which a neighbor is declared dead. Typically, this value is four times the hello interval. This value also must match to create a neighbor relationship.

OSPF neighbor states

OSPF uses a Finite State Machine (FSM) to process the neighbor state transitions between routers when certain conditions are satisfied. This process can be divided into two main phases:

- Establish neighbor adjacencies.
- Synchronize OSPF databases.

It is important to clarify that in a broadcast network, database synchronization only happens between the Designated Router (DR) and DROTHER routers, and between the Backup Designated Router (BDR) and DROTHER routers (these topics will be covered in detail in the upcoming sections). In other words, DROTHER routers only establish neighbor adjacencies and stay in the 2-WAY state (which is considered a stable state) but never synchronize databases.

As an example, consider that two core switches attempt to become OSPF neighbors over a broadcast network, as shown in the following example. Let us look at the various states in this process. Neighbor adjacencies are established in steps 1 through 5, and in the rest of the steps the OSPF database is synchronized.

1. Core-1 and Core-2 start with the DOWN neighbor state.

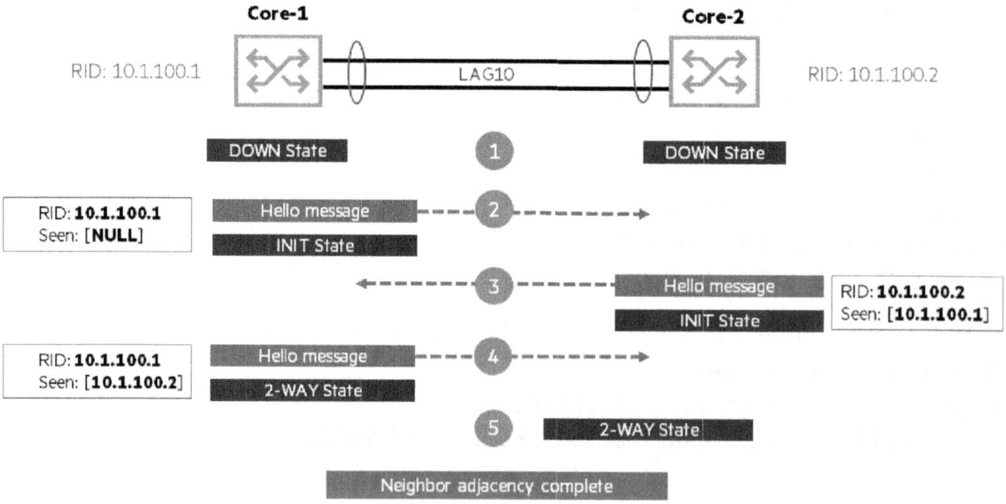

2. Core-1 transits to the INIT state when it sends the first hello message, which includes router ID = 10.1.100.1. The hello message includes a Seen field, which indicates which routers Core-1 has received hellos from. Since this is the first hello message, the field has a NULL value.

3. Core-2 receives Core-1's hello message and responds, indicating that the message has been seen and the values are compatible. Core-2 transits to the INIT state.

4. Core-1 receives the hello message and transits to the 2-WAY state. This device sends the hello message to Core-2 again, but this time, it includes both router IDs in the Seen field.

5. Core-2 receives this message and moves to the 2-WAY state.

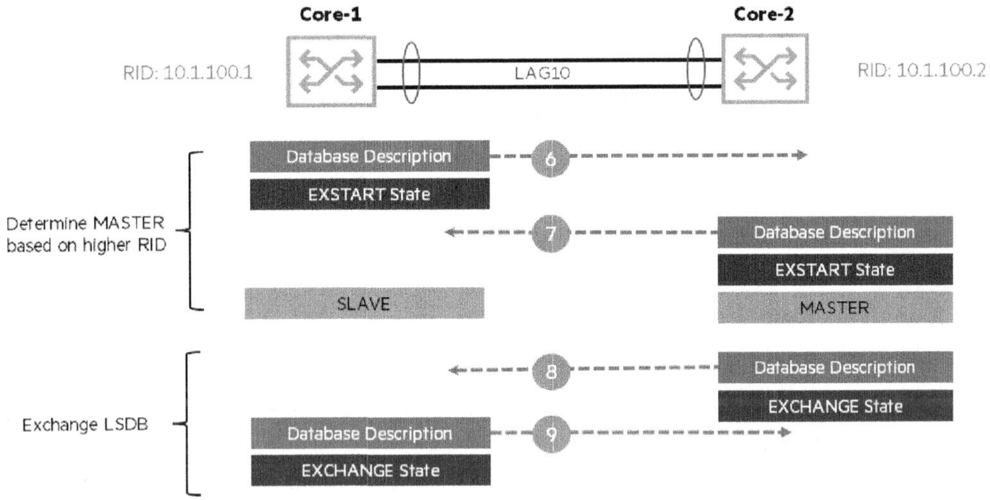

6. Core-1 initiates the synchronization process by sending a Database Description packet. The switch transits to the EXSTART neighbor state.

7. Core-2 does a similar process: it moves to the EXSTART state by sending a Database Description packet. The goal of the EXSTART state is to determine which switch will become the MASTER on the link, based on the highest router ID. The MASTER role defines only which switch will initiate the database exchange process. This role has nothing to do with DR/BDR; it is only for LSDB exchange. In this example, Core-2 has a higher router ID and becomes the MASTER.

8. Core-2 sends another Database Description packet as it transits to the EXCHANGE state. Core-2 is now sharing a summary of the contents of its LSDB with Core-1.

9. Core-1 also sends a Database Description packet and moves to the EXCHANGE state, sharing its LSDB with Core-2.

After several packets are sent and received in the EXCHANGE state, each router will have a summary of each other's entire LSDB. They compare the information received and, in the next step, request the missing information.

10. Core-1 and Core-2 request the missing information using Link State Update (LSU) packets. The other peer answers by sending the requested database information. Both switches transition to the LOADING neighbor state.

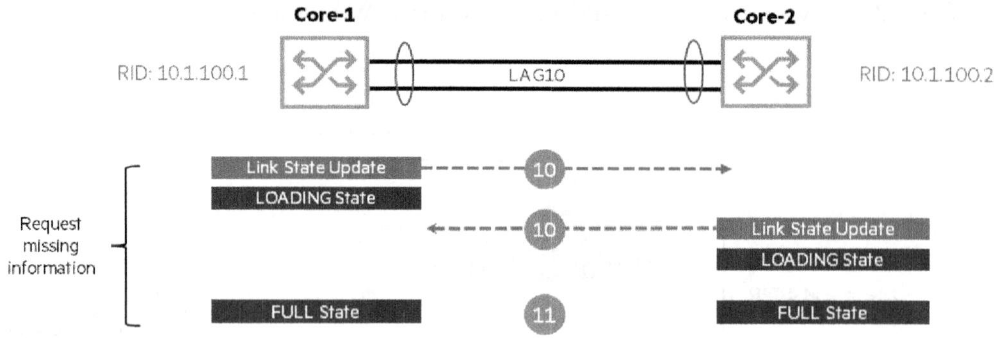

11. Core-1 and Core-2 move to the FULL state when there is no more information to be exchanged, and both devices have the same LSDB.

You can verify the OSPF neighbor state using the `show ip ospf neighbors` command.

```
Core-1# show ip ospf neighbors
Neighbor ID      Priority  State          Nbr Address      Interface
-------------------------------------------------------------------------
10.1.100.2       1         FULL/DR        10.11.0.2        LAG10
```

OSPF network types

In the context of the protocol used, the OSPF network can be classified into two types:

Point-to-point network

Only two peers are on the link. When an interface is configured to be part of a Point-to-Point Protocol (PPP) link, OSPF knows that a single neighbor device is expected on the interface. PPP serial links (not used much in networks anymore) are an example of this network type.

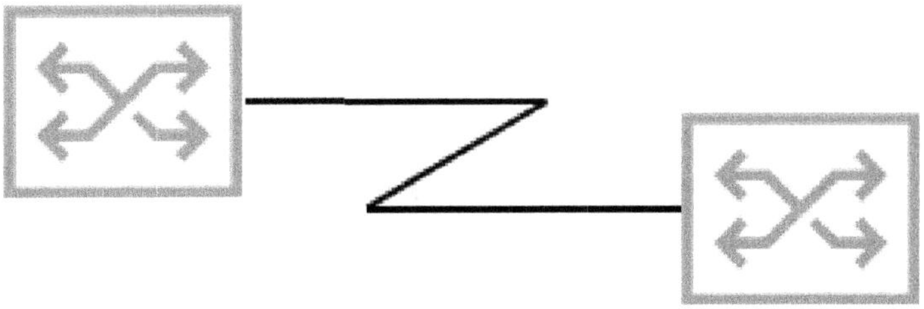

Figure 9-6: Two peers on a link

Broadcast network

Two or more peers might be on the link. When an interface is configured to be part of a broadcast network, OSPF knows that more than one neighbor might be discovered on the interface.

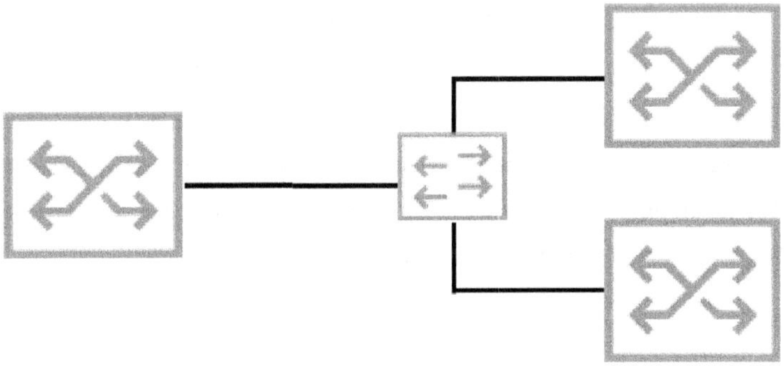

Figure 9-7: Multiple peers on a link

In AOS-CX, interfaces default to the broadcast network type, but you can modify this with the `ip ospf network` command.

```
Switch(config)# interface <INTERFACE-ID>
Switch(config-if)# ip ospf network {broadcast | point-to-point}
```

To verify the type of network that is in use, you can use the `show ip ospf interface` command. Unless you truly have multiple routers in the same broadcast domain, it is recommended to configure switch-to-switch links with the point-to-point network type.

Broadcast network scalability problem

The amount of information that can be exchanged between OSPF peers could be high in a medium-sized enterprise network. In a broadcast-type network, this could easily increase because each router can have multiple peers. This could impact router performance when hundreds or thousands of routes must be computed for each OSPF peer, as shown here.

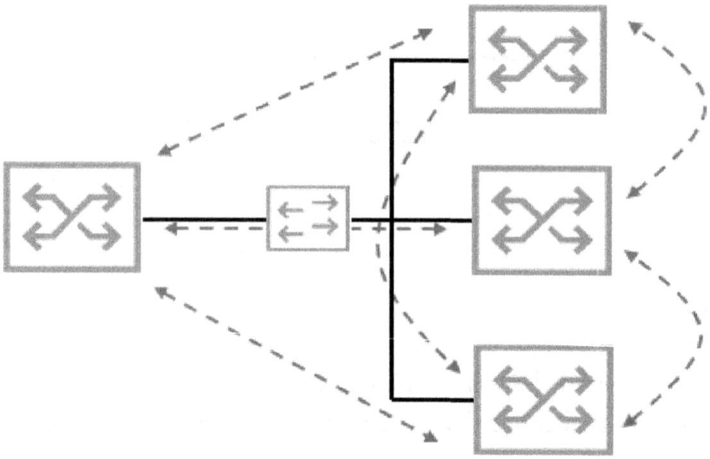

Figure 9-8: Computing routes

OSPF solves this scalability challenge by electing a Designated Router (DR) in the broadcast domain. This device maintains a complete full neighbor state with the rest of the devices (which implies that databases are exchanged with the peers). However, the non-designated routers do not exchange database information with each other. This helps to reduce the amount of information that each router in the domain must process, as shown here.

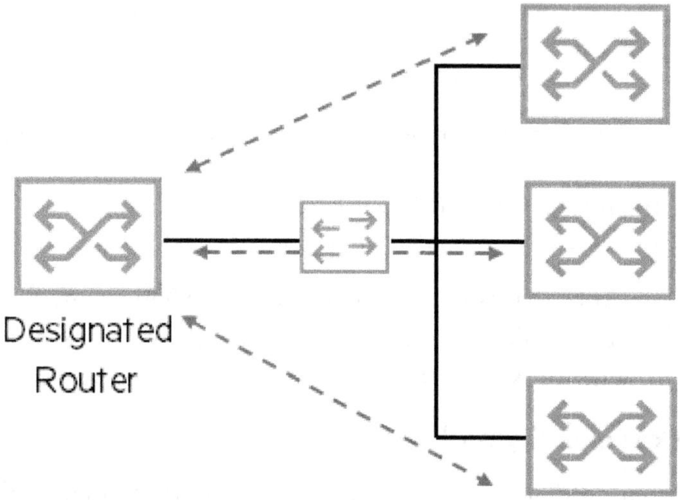

Figure 9-9: Reduced information needed by routers

To maintain high availability, you can elect a Backup Designated Router (BDR) to avoid a single point of failure. This device also maintains a full state with all devices in the broadcast network. However, it only advertises information when the primary DR is no longer available.

Designated Router, Backup Designated Router, and DROTHER

In a broadcast network, OSPF routers expect to form neighbor relationships with more than one peer. This situation could potentially be inefficient; when there is a topology change, LSAs can be flooded and reflooded in the broadcast domain. The solution is to elect a single point of contact, a Designated Router (DR), which helps to coordinate peer interactions in the broadcast domain. It forms full adjacencies with all other OSPF peers in the broadcast domain so that it can receive and flood LSAs. A Backup Designated Router (BDR) is elected to take over if the DR fails. The BDR also forms a full adjacency with the DR and all other peers in the broadcast domain.

Any router not elected to be the DR or BDR is labeled as DROTHER. These routers only form a full adjacency with the DR and the BDR. This is where the efficiency is realized. Whether there are three routers or 30 routers in the broadcast domain, DROTHERs need only form adjacencies with two peers—the DR and BDR.

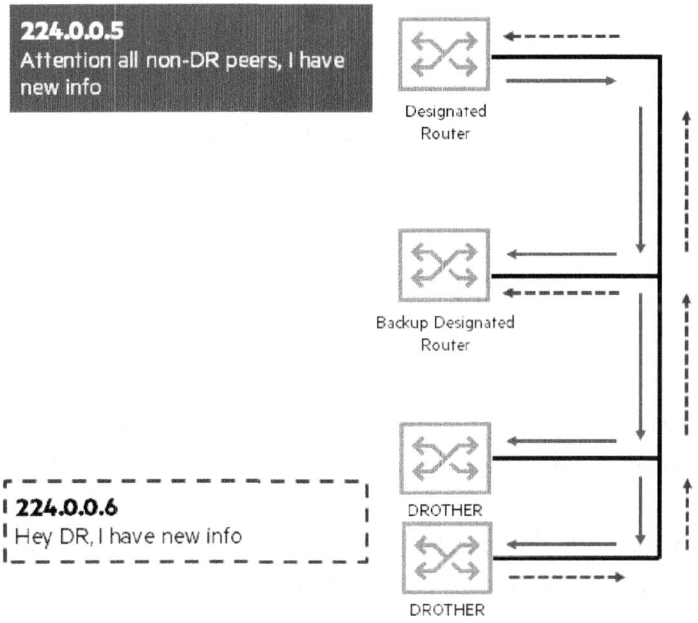

Figure 9-10: DR, BDR and DROTHER exchanging information

In the above example, one of the DROTHERs detects a topology change on one of its interfaces. It need not communicate to every other router in the broadcast domain, only the DR and BDR. So, it sends a multicast LSA using destination IP address 224.0.0.6, the reserved address that says, "Attention DR and BDR." The DR is functional, so it receives this packet. It updates its topology database and informs all non-DRs in the network with a multicast LSA to 224.0.0.5: "Attention all DROTHERs, I have new information".

The DR election is permanent. If the DR fails and BDR takes its place as stability takes precedence.

Designated router election

DR and BDR election is based on a priority value assigned to an interface; the highest priority value wins the election. In case of a tie, the router with the higher router ID becomes the DR. AOS-CX follows certain rules related to the priority value, as follows:

- The default priority value is 1.
- Valid priority values are 0-255.
- A priority value of 0 means that the router does not participate in DR election.

The configuration must be performed at the interface level.

```
Switch(config)# interface <INTERFACE-ID>
Switch(config-if)# ip ospf priority <PRIORITY-VALUE>
```

To verify the priority value and the elected DR, use the `show ip ospf interface` or `show ip ospf neighbors` commands.

```
Switch# show ip ospf interface
Transit delay 1 sec, Router priority 1
Designated Router IP: 10.0.13.2
Backup Designated Router IP: 10.0.13.1c
```

OSPF scalability

By now, you might have noticed the potential scalability challenge with OSPF. You have learned that every single router sends LSAs that advertise nearly all the contents of their LSDB to every other router. In a scenario with hundreds or thousand of routers, these LSA packets can begin to consume significant bandwidth. The LSDB (topology database) can grow quite large, consuming memory resources on each router. The SPF algorithm must then run on this very large database, consuming CPU cycles. This is not scalable. The solution is the OSPF hierarchy, implemented by splitting the network up into areas as follows.

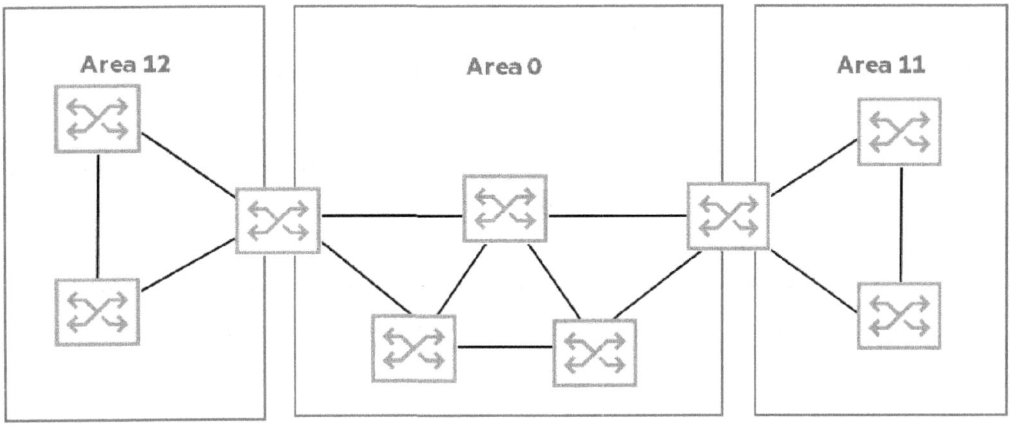

Figure 9-11: A large OSPF network split up into smaller areas

This reduces the size of the topology database each router maintains, decreasing the number of LSAs and other traffic sent on the network.

OSPF area

An OSPF area is a group of OSPF routers that share the same link state database. All routers must be part of an area. When you split a large network into separate areas, you reduce the size of the LSDB in each router, lower CPU utilization, and increase overall network stability. This is because each router in an area must only maintain the topology for that area.

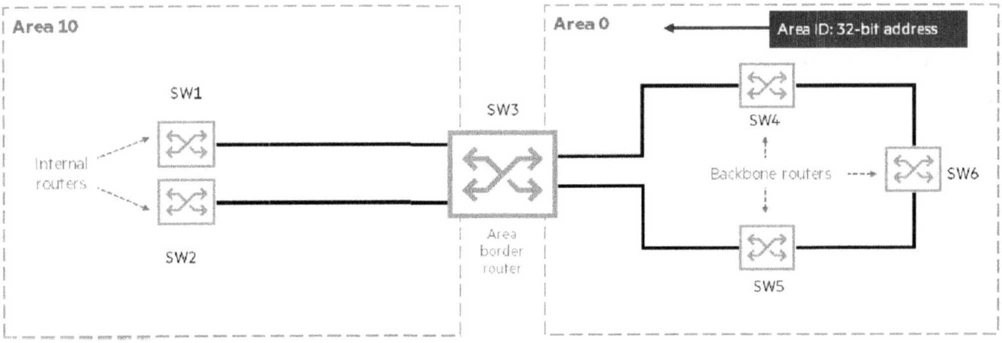

Figure 9-12: OSPF area

In this example, SW1, SW2, and SW3 need not learn about the entire topology. They only need to know about Area 10's topology. Routers SW1 and SW2 are called "internal routers" as all their interfaces are in a single area. If an internal router must route outside of its area, it simply forwards packets to its Area Border Router (ABR). An ABR is a router connected to two or more areas. SW3 is the ABR for Area 10, with two interfaces in area 10 and two interfaces in Area 0. By definition, all areas must connect directly to the special backbone Area 0. This is the most important area in the hierarchy, and so it should have a redundant design. Area 10 cannot connect directly to any other area that is not 0. Communication between two non-backbone areas must go via the backbone Area 0. Routers with interfaces in Area 0 are called "backbone routers." Routers SW4, SW5, and SW6 have all interfaces in a single area, so they are also internal routers. These can be called "internal backbone routers."

You can assign interfaces to an area by assigning them to an area ID, a 32-bit address that can be written in dotted-decimal notation (Area 0.0.0.0), or in decimal notation (Area 0, as shown in the example). AOS-CX supports both notations.

This guide focuses only on single-area designs, in which all router interfaces are in Area 0. You can learn more about hierarchical OSPF solutions in the advanced courses.

OSPF LSA Type 1 (LSA 1)

OSPF routers generate different types of Link State Advertisements (LSAs), each with a different purpose and scope in the OSPF routing domain. Each LSA includes the link state identifier, which describes the portion of the network that a router announces. Let us learn about these LSA types, starting with LSA Type 1.

The purpose of LSA Type 1 is for a router to announce itself, which is why they are called "Router LSAs." You might walk into a room and say, "Hello everyone, my name is James Bond." Similarly, routers use a Type 1 LSA to say, "Hello everyone, I am RID 10.1.100.1, and I have three interfaces that are functional and are participating in this OSPF area." It is important to remember that the information shared on LSA Type 1 depends on the link type as listed in the following example.

Data exchanged in the Link ID

Link type	Link ID	Link data
Point-to-Point	Neighbor router ID	IP address
Transit	IP address of DR	IP address
Stub	IP network	IP address mask

Link State ID describes the portion of the network that a router announces

In addition to defining the network type, OSPF also defines the link type, which are two different concepts. Link types are primarily used to describe the interfaces or neighbors of an OSPF router. The different types of links are listed as follows.

- **Stub Link:** Used when OSPF is enabled on an interface and no OSPF neighbor exists on the interface. For example, a loopback interface is considered a stub network.
- **Transit Link:** Used in a broadcast network with two or more OSPF neighbors.
- **Point-to-Point Link:** Used in point-to-point networks. Only one OSPF neighbor is expected in the link.

The link type is a consequence of the setup of the network ID and the number of neighbors that are on the link. When the network type is set to point-to-point, then the router expects a single device on the link. When the network type is set to broadcast, then the stub link (for no neighbor) or transit link (one or more neighbors) type is used.

In AOS-CX, you can verify this information by using the `show ip ospf lsdb` command.

LSA 1 for different link types

We are going to look at the information shared by Data Center Switch's LSA Type 1 advertisements when different link types are configured.

Stub link

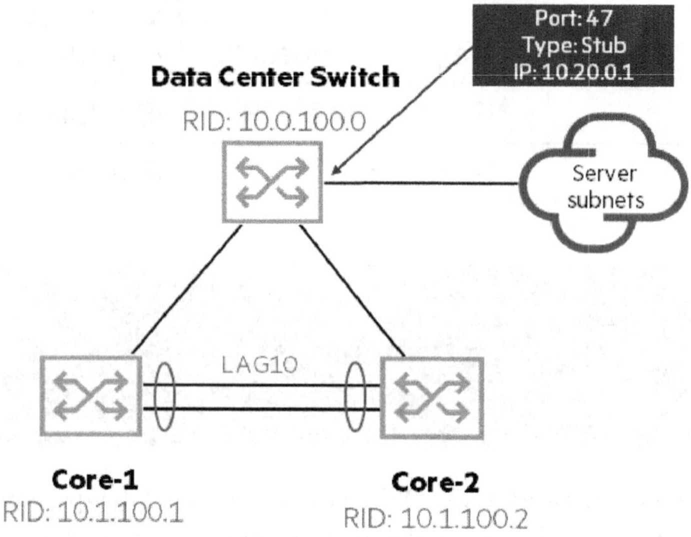

Port 47

Link Type: Stub

RID: 10.0.100.0

Link ID: 10.20.0.1

Data: 255.255.240.0 – Subnet mask

First, the stub link type includes the subnet and the mask. This information is enough to run the SPF algorithm locally to reach the destination.

Point-to-point link

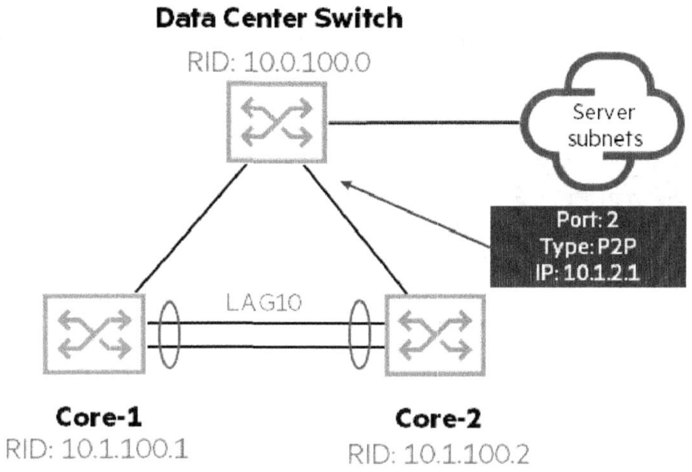

Link Type: Point-to-point

RID: 10.0.100.0

Link ID: 10.1.100.2 – RID of the peer

Data: 10.1.2.1 – Interface IP address

The point-to-point interface includes the router ID of the peer and the local interface used. From Core-1's perspective, the link details are not complete; Core-1 needs the data of the other peer on the link. This information will be known when Core-1 receives the LSA Type 1 from Core-2. When we receive the information from both parties, then Core-1 can run the SPF algorithm.

Broadcast link

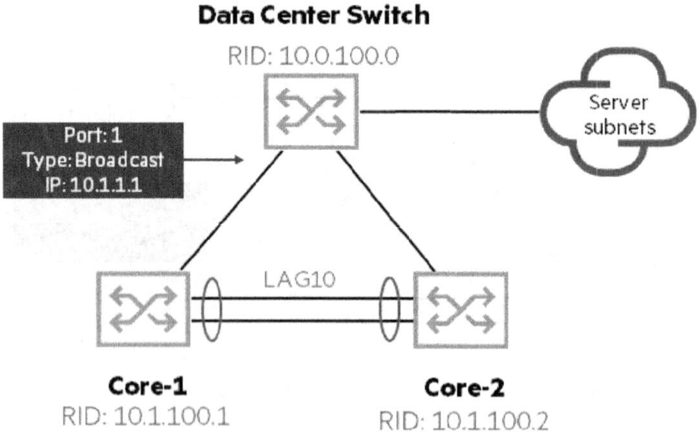

Link Type: Broadcast

RID: 10.0.100.0

Link ID: 10.1.1.1 – Interface of DR

Data: 10.1.1.1 – Interface IP address

The broadcast interface will include only the IP address of the router ID, but in this case, no subnet information will be included using LSA Type 1. To overcome this problem, we have LSA Type 2, which we will be discussed in the next section.

OSPF LSA Type 2 (LSA 2)

LSA Type 2 is used when there is a broadcast network-type link, which has an elected DR and BDR. Routers on point-to-point network types do not generate a Type 2 LSA. In this case, the link state ID is the IP address of the Designated Router. These LSAs are known as Network LSAs because each router advertises the networks for which it is the DR.

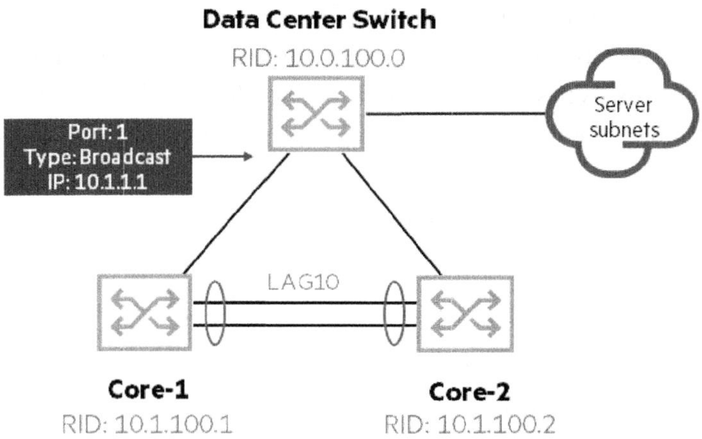

The following image shows a packet capture using Wireshark where the netmask is included in the LSA. This information helps to run the SPF algorithm. The information that was missing in Type 1 is now known.

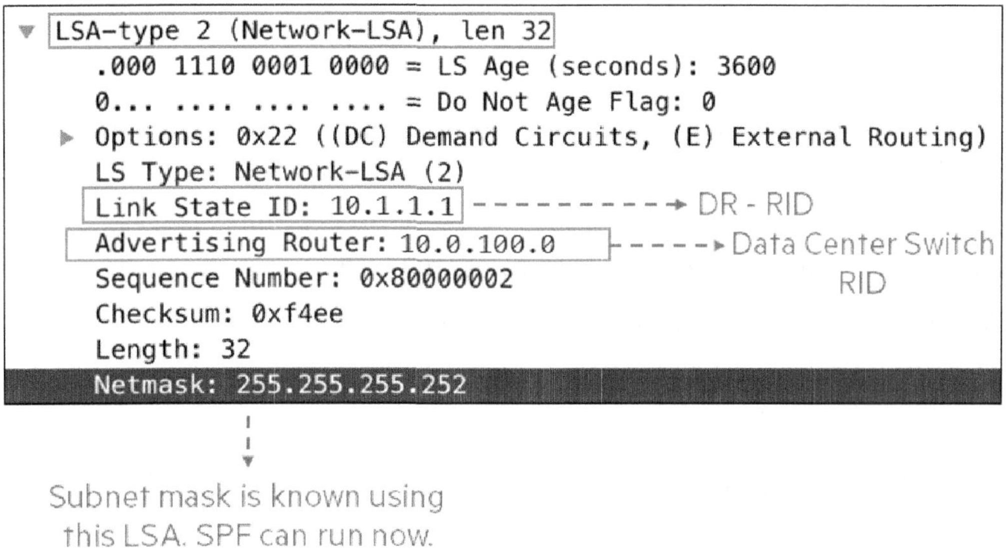

When you run the `show ip ospf lsdb` command, the Network Link State Advertisement section shows you a list of DRs.

Path selection

Let us now learn how a path is selected to route the traffic. After all routers have successfully exchanged LSAs and LSUs, they all have an identical LSDB, a topology database. Remember, the LSDB is a list of every link, every router, and how those routers and links are interconnected. It is a list of every path. Now the routers must run the Shortest Path First (SPF) algorithm (Dijkstra's algorithm) to find the best paths to each destination subnet. The best path is the one with the lowest cost, and by default, cost is based on bandwidth. Therefore, if there are multiple paths to a single subnet, OSPF chooses the path with the lowest cumulative cost, which is the fastest path.

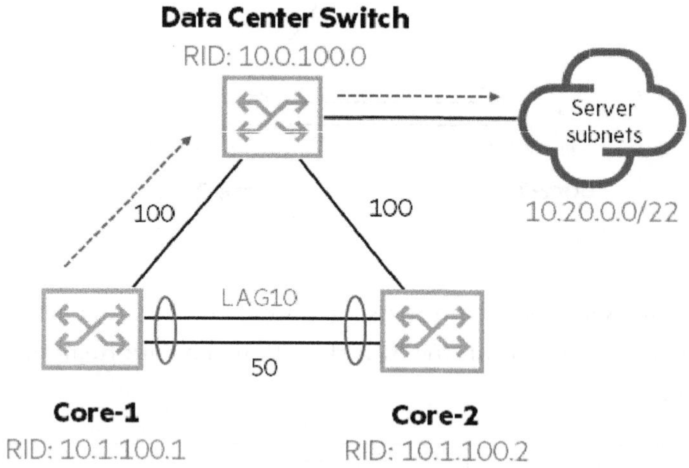

Figure 9-13: Path selection example

Consider the topology shown above. Core-1 has two paths to reach destination subnet 10.20.0.0/22. To determine each path's cost, simply add the cost values indicated for each link. Here is what you get:

Using Data Center Switch as the next hop, the cost is 100 = 100.

Using Core-2 as the next hop, the cost is 100+50 = 150.

Obviously, using Data Center Switch results in the lowest cost, and so that path is added to the route table.

Table 9-1: OSPF path selection

Route	Cumulative cost	Best path
Core 1 - Data Center Switch	100	Yes
Core 1 - Core 2 - Data Center Switch	150	No

OSPF convergence

In an OSPF routing domain, when a topology changes occurs, then the OSPF database must be updated. There are two components to OSPF routing convergence:

- Detecting topology changes
- Recalculating routes

Topology change detection is supported in two ways by OSPF. The first, and quickest, is a failure or change of status on the physical interface. The second is a timeout of the OSPF hello timer. An OSPF neighbor is deemed to have failed if the time to wait for a hello packet exceeds the dead timer, which defaults to four times the value of the hello timer. The default hello timer is 10 seconds, so the default dead timer is 40 seconds.

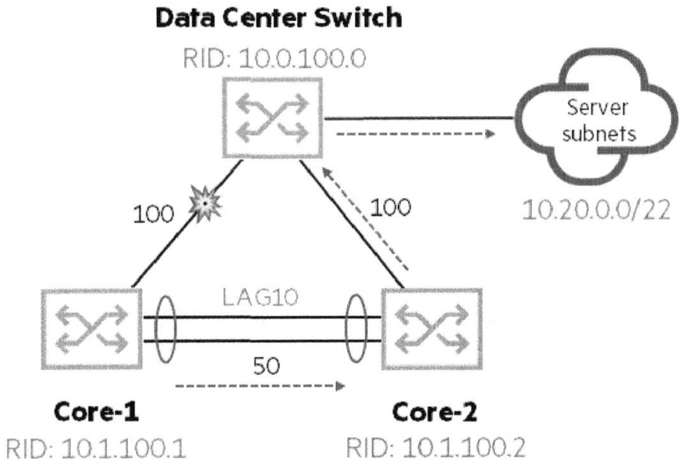

Figure 9-14: OSPF convergence

When a change is detected, an LSA is sent to all routers in the OSPF area to signal the topology change. In this example, the link between Core-1 and Data Center Switch has failed. Data Center Switch and Core-1 detect this outage, and their links go from an UP state to a DOWN state. Thus, Data Center Switch and Core-1 originate the topology change LSA. Then, they run the SPF algorithm to calculate their new best paths to any affected network, such as 10.20.0.0/22. Data Center Switch will not change anything because the best path is directly connected to it. Core-1, on the other hand, will recalculate its best path. In this case, it will use the path using Core-2. Ultimately, Core-2 receives LSAs about the change, but no recalculation is needed from its point of view. Each router performs route recalculation after a failure has been detected. This causes all routers to recalculate their routes using Dijkstra's (SPF) algorithm.

Passive interfaces

OSPF configuration involves enabling the protocol on logical and physical interfaces. Thus, the router can generate LSAs to advertise subnets to other routers. This implies that the router sends periodic hello messages on all OSPF-enabled interfaces.

In some cases, this is not desired. The most common example is when only hosts exist on a subnet and the router constantly sends OSPF packets on a network where nobody cares. Hosts do not respond to OSPF multicast packets; it simply wastes processing cycles and bandwidth on the link. If bad actors are on the link, they could learn information about your network and potentially mount an attack.

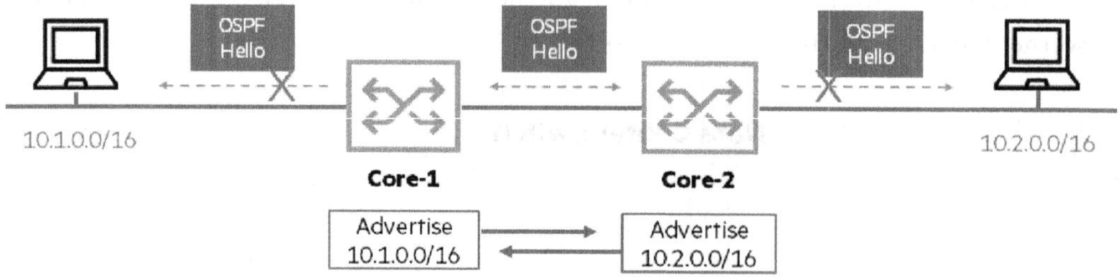

Figure 9-15: Passive interfaces

The solution is to configure these host-facing router interfaces as passive. When an interface is passive, it stops sending and receiving OSPF packets out that interface. However, it continues to advertise about that interface to other routes. In the above example, Core-1 does not send hello packets out its 10.1.0.0/16 network, but it will tell Core-2 about that network. Core-2 has a similar arrangement, as relates to the 10.2.0.0/16 link.

In AOS-CX, you can use the `ip ospf passive` command in the interface context to enable a passive interface.

```
Switch(config)# interface <INTERFACE-ID>
Switch(config-if)# ip ospf passive
```

Reference bandwidth

The cost associated for each interface is calculated using this formula:

OSPF cost value = (Reference Bandwidth)/(Interface Bandwidth)

AOS-CX uses a default reference value of 100,000 Mbps. You can verify this value using the `show ip ospf` command.

```
Switch # show ip ospf | include Bandwidth
Reference Bandwidth: 100000 Mbps
```

The result of the OSPF cost formula can be displayed using the `show ip ospf interface` command. Notice that a 10 Gbps port has assigned the cost of 10.

```
Switch# show ip ospf interface
IP address 10.0.13.1/24, Process ID 1 VRF default, area 0.0.0.0
State Backup-dr, Status up, Network type Broadcast
Link Speed: 10000 Mbps
Cost Configured NA, Calculated 10
Transit delay 1 sec, Router priority 1
```

Configuring cost value

AOS-CX allows you to modify the cost of an interface in two different ways.

Modifying the reference bandwidth value is the first way. This method is applied to the entire OSPF process, which means that will affect all interfaces.

```
Switch(config)# router ospf <PROCESS-ID>
Switch(config-ospf-1)# reference-bandwidth <1-4000000>
```

Modifying the cost associated with an interface is the second way. This approach allows you to only modify the cost value to a specific interface. When this command is applied, the router no longer considers the cost formula but instead simply uses the manual value that was entered.

```
Switch(config)# interface <INTERFACE-ID>
Switch(config-if)# ip ospf cost <COST-VALUE>
```

Configuring OSPF in AOS-CX

The commands to configure OSPF in AOS-CX are listed as follows.

Enable OSPF process

AOS-CX is capable of running an OSPF process per VRF. The `router ospf <PROCESS-ID>` command is used to enable the OSPF process and assign a process value to it, a number between 1 and 63.

```
Switch(config)# router ospf <PROCESS-ID>
```

Configure the router ID

```
Switch(config-ospf-1)# router-id <A.B.C.D>
```

Create the OSPF area ID

AOS-CX allows you to configure the area ID as a dotted-decimal value or decimal format. For example, 0.0.0.0 is equivalent to 0. Dotted-decimal values can range from 0.0.0.0 to 255.255.255.255. Decimal values can range from 0 to 4,294,967,295.

```
Switch(config-ospf-1)# area <AREA-ID>
```

Enable OSPF in the interface

```
Switch(config)# interface <INTERFACE-ID>
Switch(config-if)# ip ospf <PROCESS-ID> area <AREA-ID>
```

Set the network type

```
Switch(config)# interface <INTERFACE-ID>
Switch(config-if)# ip ospf network {broadcast | point-to-point}
```

Set the OSPF cost associated to the interface (optional)

```
Switch(config)# interface <INTERFACE-ID>
Switch(config-if)# ip ospf cost <COST-VALUE>
```

Learning check

Check your understanding of this material by answering the following questions.

Question 1

Which of the following are NOT a primary characteristics of OSPF? (Select two)

- ☐ Dynamic routing
- ☐ Scalability
- ☐ Fast convergence
- ☐ Distance vector routing
- ☐ Link State routing

Question 2

What IP protocol number is associated with OSPF?

- ○ 6
- ○ 17
- ○ 89
- ○ 123
- ○ 255

Question 3

What type of network is characterized by having only two peers on the link?

- ○ Point-to-point
- ○ Ethernet
- ○ Non-Broadcast Multiple Access (NBMA)
- ○ Broadcast
- ○ Token ring

10 Quality of Service

In the dynamic landscape of network management, ensuring optimal performance and resource allocation is paramount. In this chapter, you are going evaluate the importance of Quality of Service (QoS) fundamentals, principles, and techniques for prioritizing and controlling traffic.

OBJECTIVES

After completing this chapter, you should be able to:

- ✓ Define Quality of Service (QoS) and its significance in network traffic management.
- ✓ Analyze the concept of packet classification.
- ✓ Delineate queuing mechanisms.

Quality of service overview

QoS provides special network treatment to specific traffic.

QoS enables networks to provide better or special service to a set of users and/or applications, perhaps to the detriment of other users and/or applications. With QoS, you are trying to make the best possible use of available bandwidth.

Different traffic types compete for the same network resources such as bandwidth and fast processing. Without QoS, packets are served as they arrive, typically first-in first out (FIFO). Your network management traffic may be more important to you than some users watching videos during their lunch break, but, by default, there is not differentiation between this traffic.

On the highway, an ambulance gets higher priority than other vehicles. Likewise, QoS is not being fair when dividing the network resources, but rather selecting certain traffic types for improved service, perhaps at the expense of other traffic. QoS does not make the road wider, it just decides who goes first and consequently who must wait.

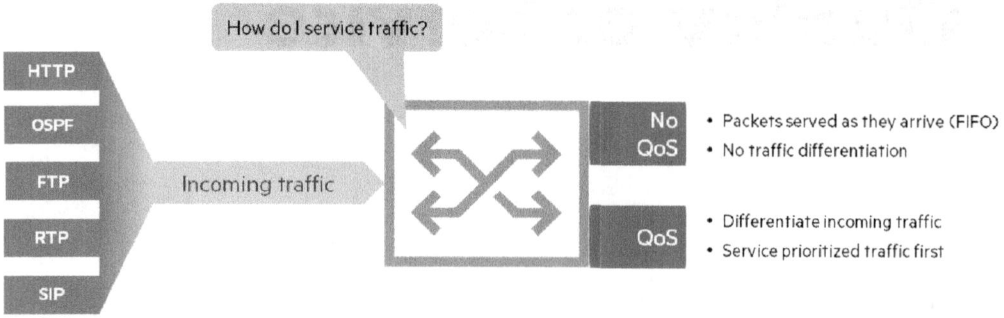

Figure 10-1: QoS overview

Understand that QoS is not a standalone service or product, but rather a concept that supports a framework of protocols, techniques, and network attributes that span across the infrastructure.

QoS introduction

Quality of Service refers to a set of techniques and mechanisms used to manage and prioritize network traffic, ensuring that certain applications or users receive adequate resources and performance levels according to predefined criteria. QoS aims to optimize network performance, reduce latency, minimize packet loss, and improve the overall user experience.

Benefits of QoS

Enhanced user experience: QoS ensures that critical applications, such as VoIP, video conferencing, and real-time data transmission, receive priority treatment, resulting in improved performance and quality for end-users.

Efficient resource utilization: By prioritizing traffic and allocating resources based on application requirements and business priorities, QoS helps maximize network efficiency and bandwidth utilization.

Reduced the impact of network congestion: QoS mechanisms such as traffic shaping and rate limiting prevent network congestion by regulating the flow of traffic and ensuring that bandwidth-intensive applications do not monopolize available resources. Like in a traffic jam, we want ambulances, firetrucks, and police cars to be able to move... So we can reserve a lane for them.

Support for mission-critical applications: QoS enables organizations to prioritize traffic for mission-critical applications and services, ensuring reliable and consistent performance even during periods of high network activity.

Compliance with service level agreements (SLAs): QoS allows service providers to meet SLA requirements by guaranteeing certain levels of performance, availability, and reliability for their customers.

The difference between the benefits and goals of QoS lies in their focus and perspective.

Goals of QoS:

Prioritizing critical traffic: QoS aims to prioritize network traffic based on its importance or criticality, ensuring that mission-critical applications—such as VoIP, videoconferencing, and real-time data transmission—receive priority treatment and optimal performance, resulting in improved performance and quality for end users.

Minimizing latency: QoS seeks to minimize latency or delay in network communications, particularly for time-sensitive applications that require real-time or low latency data transmission. By reducing latency, QoS improves responsiveness and user experience for interactive applications.

Reducing packet loss: QoS endeavors to minimize packet loss by implementing mechanisms to handle congestion and prevent network resources from becoming overwhelmed. By reducing packet loss, QoS helps maintain data integrity and ensures the reliable delivery of information across the network.

Maximizing bandwidth utilization: QoS aims to maximize the efficient use of available network bandwidth by allocating resources based on application requirements and business priorities. By optimizing bandwidth utilization, QoS helps prevent network congestion and ensures fair access to resources for all users.

Ensuring fairness and equity: QoS seeks to ensure fairness and equity in resource allocation by implementing policies that prioritize traffic based on predefined criteria, such as type, user identity, or Service Level Agreements (SLAs). By enforcing fairness, QoS helps prevent certain applications or users from monopolizing network resources at the expense of others.

Supporting SLAs: QoS aims to meet the requirements and commitments outlined in SLAs between service providers and customers. By guaranteeing certain levels of performance, availability, and reliability, QoS helps organizations comply with SLAs and maintain high-quality service delivery standards.

QoS principles

The principles of QoS revolve around the systematic management and prioritization of network traffic to ensure optimal performance, reliability, and user experience. The core principles of QoS include:

1. Traffic classification, marking, and prioritization:
   - Classification: Traffic categorized based on various criteria such as application type, source/destination addresses, or user-defined policies. This helps in identifying and grouping similar types of traffic.
   - Marking: Assigns labels or tags to traffic packets to indicate their priority level within the network. This helps network devices recognize and treat packets according to their priority.
   - Prioritization: Once classified, traffic is assigned different priority levels. Critical applications like VoIP and videoconferencing are given higher priority over less critical traffic, such as email or file transfers, to ensure timely and reliable delivery.
2. Traffic shaping and policing:
   - Traffic shaping: This technique controls the flow of data into the network to ensure a smooth and consistent transmission rate. By buffering and scheduling packets, traffic shaping prevents bursts of traffic from overwhelming network resources.
   - Traffic policing: Policing monitors traffic and enforces predefined bandwidth limits. Excess traffic is either dropped or marked for lower priority, ensuring that no single user or application consumes more than its fair share of resources.
3. Congestion management:
   - Queuing mechanisms: Different queuing algorithms like Priority Queuing (PQ), Weighted Fair Queuing (WFQ), and Class-Based Weighted Fair Queuing (CBWFQ) manage how packets are stored and transmitted from network buffers. These mechanisms ensure that higher-priority traffic is transmitted first.
   - Buffer management: Efficient buffer management helps in handling temporary surges in traffic, reducing packet loss, and maintaining quality of service.
4. Resource reservation:
   - Reservation protocols: Protocols like Resource Reservation Protocol (RSVP) allow applications to reserve bandwidth across a network path. This ensures that the required resources are available when needed, providing guarantees for latency, jitter, and packet loss.
5. Policy-based management:

- Policy enforcement: Network policies define how different types of traffic are handled based on organizational requirements. Policies are enforced through network devices, ensuring consistent QoS across the network.

6. Monitoring and adjustment:
   - Performance monitoring: Continuous monitoring of network performance metrics like latency, jitter, and packet loss helps in assessing the effectiveness of QoS policies.
   - Dynamic adjustment: Based on performance data, QoS configurations can be dynamically adjusted to adapt to changing network conditions and maintain optimal performance.

By adhering to these principles, QoS ensures that critical applications receive the necessary bandwidth and low latency they require, while efficiently managing overall network resources to prevent congestion and maintain high performance across all types of network traffic.

Traffic loss, latency, and jitter

Consider your needs and take steps to protect the higher-priority classes of traffic from the ill effects of congestion: loss, latency, and jitter. Each application has different needs and behaviors when dealing with those effects. Since voice applications tend to be the most sensitive to all three of these issues, we will use voice traffic to illustrate how these issues affect voice traffic as a use case.

Packet loss occurs when congestion prevents the network from delivering all packets. While voice applications can tolerate a few lost packets, too many lost packets will degrade quality and eventually drop the call.

Latency is the amount of time that passes between the sending of a transmission and its arrival at the receiving station. Within a campus network, significant latency typically occurs, not due to the transmission time for signals but because congestion causes a packet to be buffered before transmission. Latency can also happen if an endpoint's connection is disrupted by roaming or re-authentication. Latency can cause issues for voice traffic because it introduces a slight delay after a person begins to speak. This delay can cause people to talk over each other and sometimes creates an echo. The International Telecommunications Union (ITU) recommends no more than 150 ms end-to-end delay for voice traffic.

Because congestion typically occurs in unpredictable bursts, the amount of latency can vary from packet to packet within a stream. This effect is called jitter. Jitter can have a much worse impact on call quality than simple latency because real-time applications are designed to have a consistent packet rate, and jitter causes smaller and more significant gaps between packets. It can sometimes even cause packets to arrive out of order. For instance, a VoIP telephone sends a very small packet every 20-30 ms. If the interval between packets grows to 50 milliseconds (ms) due to jitter, the transmission will be unsatisfactory for many users.

VoIP is particularly sensitive to many of these issues because it is a real-time application that uses UDP. TCP-based applications can recover from packet loss because they use acknowledgments. However, resending the packets does slow down the application. UDP applications do not use acknowledgments and cannot recover lost packets. In a real-time conversation, by the time the packet could be recovered, it is too late for it to be useful.

You should also be aware that some latency is inevitable. While the speed of light in a vacuum is 186,282 miles per second (299,792 km per second), signals move through fiber or copper cable at approximately 125,000 miles per second (200,000 km per second). Consequently, a fiber network stretching halfway around the Earth introduces a delay of approximately 70 milliseconds under the best of circumstances, with an additional delay introduced by the physical transport medium. However, within a campus, the latency introduced by the physical media is negligible. Instead, you need to worry about the latency introduced by congestion.

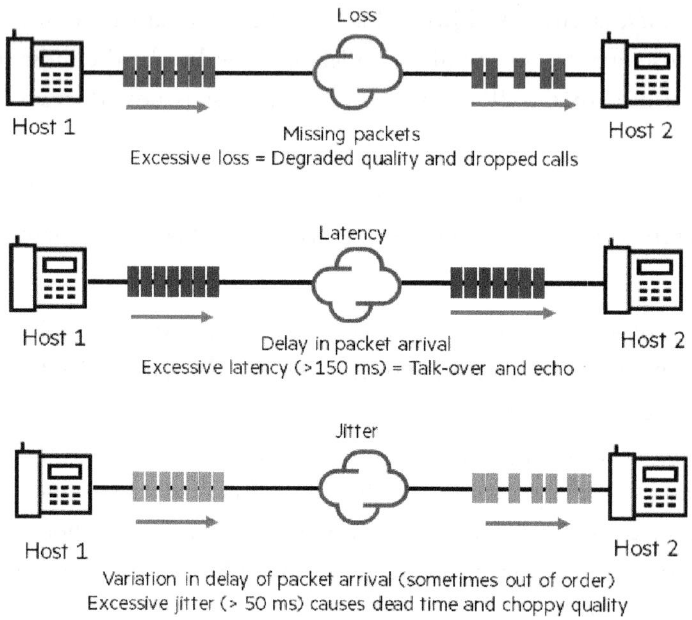

Figure 10-2: Traffic loss, latency, and jitter

Here are some key terms related to QoS.

Bandwidth

Definition: Bandwidth refers to the maximum data transfer rate of a network or internet connection. It represents the capacity of the network to transmit data over a specific period.

Importance: Adequate bandwidth is essential for ensuring that all applications have sufficient capacity to operate efficiently. Insufficient bandwidth can lead to network congestions, which can degrade the performance of applications, especially those requiring high data rates such as video streaming or online gaming.

QoS mechanisms: Bandwidth management techniques, such as traffic shaping and rate limiting, are used to allocate and prioritize bandwidth among different types of traffic to ensure that critical applications receive the necessary resources.

Delay

Definition: Delay, or latency, is the time it takes for a data packet to travel from its source to a destination across a network.

Importance: Low latency is crucial for real-time applications like VoIP and webinars, where delays can significantly impact the user experience. High latency can result in noticeable lag, making real-time communication difficult.

QoS mechanisms: Techniques such as prioritization and the use of a low-latency queuing mechanism help to minimize delay for critical applications.

Jitter

Definition: Jitter refers to the variation in packet arrival times. It is the measure of the variability in latency between successive packets.

Importance: Consistent latency is essential for the smooth delivery of real-time applications like videoconferencing and VoIP. High jitter can cause packets to arrive out of order, leading to issues such as choppy audio or video.

QoS mechanisms: Buffering, traffic shaping, and proper sequencing techniques are employed to reduce jitter and ensure that packets are delivered in a consistent and predictable manner.

Packet loss

Definition: Packet loss occurs when one or more packets of data fail to reach their destination. It is typically expressed as a percentage of packets lost with respect to packets sent.

Importance: Packet loss can severely impact the quality of applications, especially those that are sensitive to data loss such as video streaming, VoIP, and online gaming. Lost packets can result in gaps in audio or video streams and can degrade the overall application performance.

QoS mechanisms: Techniques such as error corrections, retransmission strategies, and adequate network provisioning help to minimize packet loss and maintain the integrity of data transmission.

By effectively managing these components, QoS ensures that network resources are utilized efficiently, and critical applications receive the performance they require.

Traffic prioritization

Traffic prioritization enables the network infrastructure to place different types of traffic in different queues and then preferentially forward traffic in higher-priority queues. In this example, one IP phone device sends voice traffic, and a workstation sends data traffic. Each infrastructure device in the traffic's path needs to prioritize the voice traffic to minimize loss, latency, and jitter.

You will learn about several technologies that help network devices to recognize traffic's priority and communicate that priority to other devices. These technologies include 802.1p and Differentiated Services (DSCP).

You can break down the communications into several segments. It is important to consider how traffic is classified and handled at each segment of its journey to ensure the proper QoS.

You need to consider each link in the data path. For example, switch-to-switch links are more vulnerable to congestion, as they are typically oversubscribed with multiple downstream devices sharing the links' bandwidth. You should remember that every device in the path can affect the QoS by either correctly or incorrectly prioritizing the traffic.

Finally, remember to consider traffic in both directions of a conversation and ensure that return traffic is properly classified and prioritized.

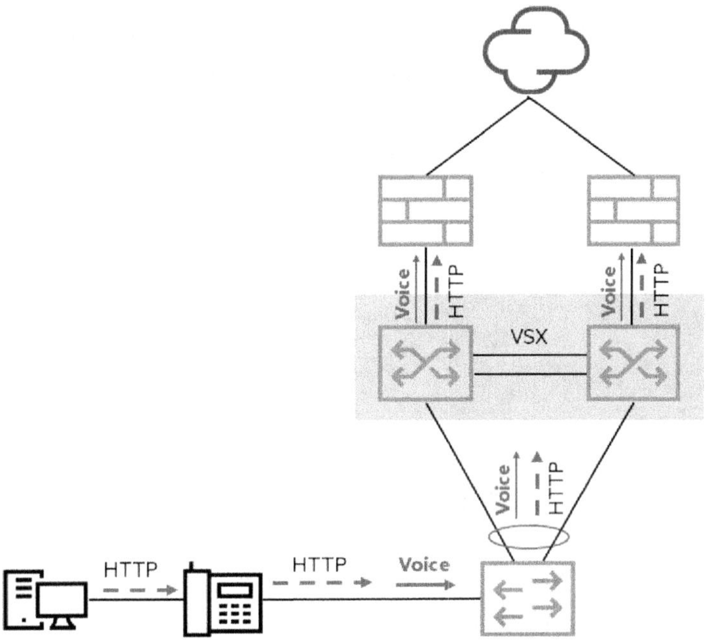

Figure 10-3: Traffic prioritization

Campus networks are often planned to have some level of over-subscription on the uplinks. It happens because if all edge devices connected to a switch were operating at near-full capacity, you would need to plan the uplink bandwidth to match the bandwidth provided to the switch ports. However, you can oversubscribe the uplinks because most wired devices do not require nearly all of their 1Gbps port.

The following formula calculates the over-subscription ratio: The number of switch ports times port bandwidth divided by uplink bandwidth equals over-subscription.

For example, an access layer switch has 48 Gigabit edge ports and 2 Gigabit uplinks. The over-subscription ratio is 48 times 1 Gbps divided by 2 Gbps, or 24. This ratio is usually reported in this format, 24:1.

This ratio assumes that the switch can use most of the bandwidth on the uplinks, which it can if the uplinks are combined in a link aggregation.

Application classification

Network control traffic should always have high priority, but generally devices prioritize this traffic automatically, so you do not need to plan to prioritize it. You do need to identify real-time applications, such as voice and video, which require special handling to ensure a good QoS.

You also need to classify traffic based on how critical it is to your organization's mission.

There are several common classes listed in order of priority, with the highest priority at the top. You can also see the types of applications associated with these classes. Often, it is a good idea to make a plan to provide the proper QoS to voice and interactive video traffic, as well as perhaps streaming video. Everything else receives best effort forwarding.

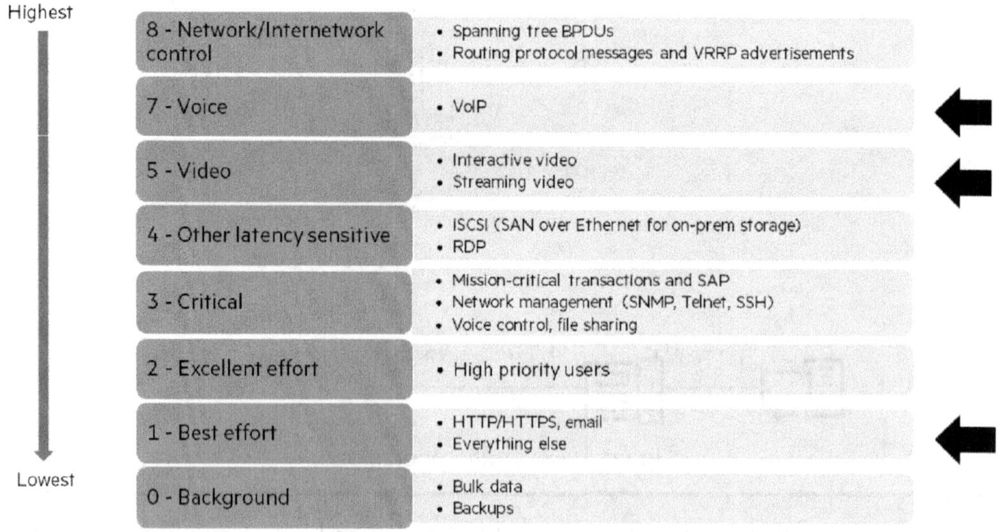

Figure 10-4: Application classification

Eventually, you will associate each traffic class with a forwarding queue. In the wireless world, devices use one or four queues, depending on their capabilities. The advantage of using fewer classes, and therefore fewer queues, is both simplicity and, for wired switches, greater buffering space per queue. In addition, the simpler the design, the easier the solution is to deploy and maintain.

As you see, network and internetwork control traffic should have the highest priority. This traffic includes control protocols transmitted by network infrastructure devices themselves. For example, internetwork control traffic includes Open Shortest Path First (OSPF) and other routing protocol messages. Virtual Router Redundancy Protocol (VRRP) advertisements provide another example of internetwork control traffic. Also, the BDPUs used by the Spanning Tree Protocol (STP) are network control traffic. All of this traffic requires the highest priority because the network cannot ensure delivery for any traffic if these protocols stop functioning correctly.

Multimedia applications, such as voice and video, require the next highest priority. Voice typically requires higher priority than video because it occurs in real time. You can further divide video into interactive video and streaming video. Interactive video might require higher priority because real-time communications require timely delivery.

Two other latency sensitive areas are iSCI and Remote Desktop Protocol. The iSCSI protocol carries SCSI communications between an initiator—typically, a server—and a storage target. This technology enables a server to mount volumes from disks hosted on a remote storage array. While you will most often find iSCSI deployed within data centers, some customers do have on-prem storage and run iSCSI traffic over the Local Area Network. The iSCSI communications require high priority to prevent excessive latency or packet loss. Remote Desktop Protocol (RDP) can also be latency sensitive.

Mission-critical traffic might include applications associated with transactional databases as well as SAP applications. Companies may define network management traffic such as SNMP, Telnet, and SSH as critical. Often, non-voice and non-video Unified Communications (UC) traffic fits in this class as well. This traffic includes voice control (the traffic for setting up a call) and file sharing.

Classifying traffic into the next two classes depends less on the application type and more on the scenario. For example, a hotel might offer a premium and a best effort service. All traffic from users who purchase the premium service should be granted critical or excellent effort.

Finally, you can identify traffic that takes a lot of bandwidth but does not require timely delivery as background traffic. For example, FTP uses TCP, which can tolerate loss and speed up and slow down based on congestion. The company might have periodic backups that need to execute but can complete anytime during the day without negative effects. You could also define applications that the customer does not care about as background traffic.

Packet classification and marking

Packet classification is a fundamental aspect of QoS that involves identifying and categorizing network packets based on predefined criteria such as application type, IP address, or protocol. This process is crucial for applying appropriate QoS policies to ensure that different types of network traffic receive the necessary levels of service and priority. When a switch receives a packet, the first thing it needs to do is classify it. This means the packet is sorted and put into classes, which you will configure. Classes are also known as forwarding classes.

Mechanisms to achieve QoS

In the classification stage, packets are analyzed and categorized based on various criteria such as the source/destination IP address, Layer 4 port numbers, or application type. The switch may classify packets as belonging to voice, video, or data traffic based on their headers and payload characteristics.

After classification, the switch will mark each packet in the Ethernet or IP header with a specific identifier. Marking after classification is a technique used to tag packets with specific identifiers that indicate their priority or handling requirements. This process helps ensure that downstream devices, such as the next hop in a network, can process the packets quickly and efficiently without having to reclassify them.

This identifier can be in the form of a CoS value or DSCP value. The marking serves as a label that tells downstream devices how to handle the packet. You may have voice packets marked with a DSCP value of 46 (expedited forwarding), video packets with a DSCP value of 34 (assured forwarding), and regular data packets with aDSCP value of 0 (best effort).

By marking packets after classification, downstream devices (or the next hop switch) do not need to perform the classification again. Instead, they simply read the marking and apply the appropriate QoS policy based on the mark.

Classification

- Recognize traffic that should be prioritized
- Assign an internal traffic class (internal forwarding priority)

Marking

- Indicates within the header how traffic should be handled – for the benefit of other switches
    - Layer 2 marking: 802.1p (CoS).
    - Layer 3 marking: IP ToS or DiffServ

Scheduling

- Place traffic in queues based on traffic class
- Allocate sufficient percentage of outbound bandwidth for high priority traffic

Layer 2 marking

In QoS configuration, externally marked traffic refers to packets that arrive at a network device with pre-assigned priority levels. These markings help network devices quickly identify the type of service each packet requires. Two common types of marking are Class of Service (CoS) and Differentiated Services Code Point (DSCP).

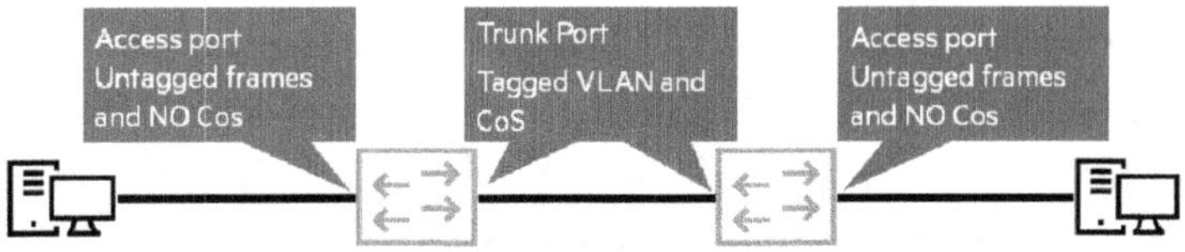

Figure 10-5: CoS marking topology

CoS is a Layer 2 (data link layer) marking used primarily in Ethernet networks. It is found in the 802.1Q VLAN tag and consists of three bits, allowing for eight different priority levels (0-7). CoS values help switches prioritize traffic within VLANs. Higher CoS values typically correspond to higher-priority traffic, ensuring that critical data flows efficiently through the network even during congestion.

CoS only works with tagged frames; therefore, the VLAN must be tagged in the link.

Many network administrators use the native VLAN for network management. As the native VLAN transmits untagged frames over the link, the switch receiving the traffic will not have the 802.Q (tag) headers to verify the CoS field.

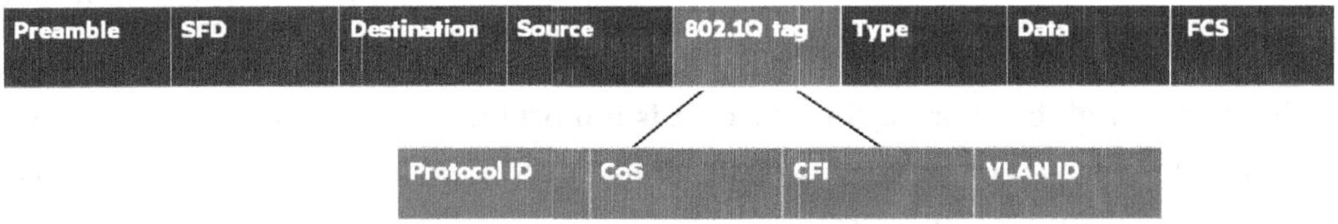

Figure 10-6: Layer 2 QoS marking: CoS

Layer 3 marking

DSCP, also known as DiffServ, is a field in the IP header used to classify and manage network traffic. It assigns a value between 0 and 63 to each packet, indicating the level of service that packet should receive. Network devices use these values to determine how to handle and prioritize the packets. For example, packets with higher DSCP values might be given higher priority, ensuring low latency and minimal jitter for time-sensitive applications like VoIP or videoconferencing.

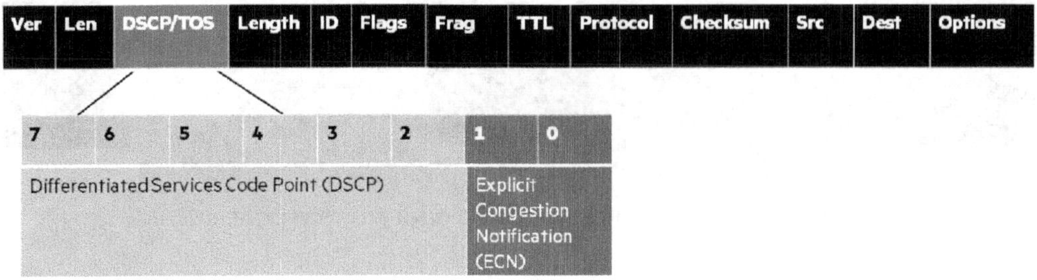

Figure 10-7: Layer 3 QoS marking: DSCP

Traffic priorities for networks can be carried in VLAN tags, using the CoS Priority Code Point (PCP), or in IP packet headers, using DSCP. Whether these priorities affect how traffic is serviced depends on how the QoS trust mode is configured on the switch. The QoS trust mode specifies how the switch assigns local priority values to ingress packets and can be set globally for all interfaces, or individually for each interface. By default, the trust mode is set to none, meaning that any QoS information in the packet (CoS or DSCP) is ignored, and local priority values are assigned from the CoS map value for code point 0. An exception to this can be configured, allowing a QoS remark to be applied to DSCP values when the trust mode is none.

Interface and global trust

When configuring QoS, interface trust and global trust are essential concepts. QoS interface trust settings determine how incoming packets are treated based on their markings, such as Differentiated Services Code Point (DSCP), Class of Service (CoS), and the interface level. This ensures that the QoS policies are applied correctly according to the packet's classification.

Global trust settings, on the other hand, define the default trust state for all interfaces on a device. By configuring global trust, network administrators can establish a consistent QoS policy across the entire network infrastructure, ensuring the uniform treatment of traffic and simplifying the QoS configuration process.

You can choose either globally or per interface. It is important to remember that the interface configuration takes precedence. For example, if the global level is set to DSCP but the interface is for CoS, CoS will be used on that interface.

> AOS-CX default is no-trust.

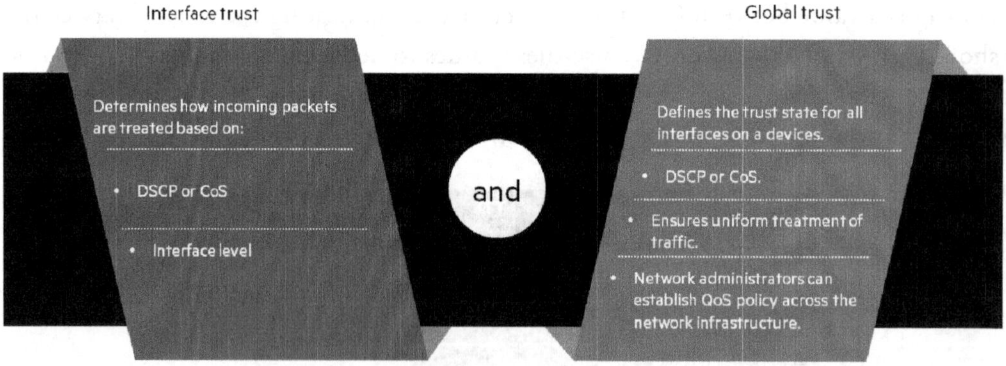

Figure 10-8: Interface trust and global trust comparison

Layer 4 classification

Traffic policing

Traffic policing supports policing of the inbound traffic. A typical application of traffic policing is to classify and remark the incoming packets based on their protocol (TCP or UDP) and port number with the proper DSCP marking, allowing the local switch and for the benefit of other switches.

Types of policy actions

The policy actions are broadly classified into the following categories:

- Remark actions
- Police actions
- Other actions

For more detailed information on traffic policing, check the *ACLs and Classifiers Policy Guide.*

Policy Based routing

Policy Based routing (PBR) lets you manipulate the path of a packet based on the various attributes of the packet. Packets that can be manipulated by PBR are packets that are already routing through the system at Layer 3, with a destination IP address that is on a network other than the packet ingress Layer 3 interface. PBR is an extension of the existing classifier policy system where traffic to be manipulated is matched by a classifier class, and policy actions to be executed on the matching traffic. Matching traffic with the same destination can be routed over different paths so that different types of traffic, such as VoIP or traffic with special security requirements, can be better managed.

An example of this is directing HTTP traffic (port 80) through a different path or applying specific treatments to it.

For more detailed information on traffic policing, check the *IP Routing Guide.*

Queuing

Queuing and queuing algorithms are critical for managing network traffic and ensuring QoS by determining the order in which packets are transmitted. Common types include First-In, First-Out (FIFO), which processes packets in the order they arrive. In this section, we are going to look at the five types of queuing algorithms and how they support QoS.

1. First-In, First-Out (FIFO)
2. Strict Priority Queuing (SPQ)
3. Deficit Weighted Round Robin (DWRR)
4. Weighted Fair Queuing (WFQ)
5. Virtual Output Queues (VoQ)

So what is a queue? A queue is used to store packets (or traffic) until they can be processed. AOS-CX switches have ingress queues and egress queues. An ingress queue stores packets until the switch can forward the data to the appropriate interface or destination. An egress queue stores packets until the switch can serialize the data onto the physical wire.

AOS-CX queues

We have shown you three queues for simplicity's sake of understanding the way the algorithm works. However, AOS-CX allows for eight queues. With QoS, every port has eight egress queues, numbered 0 to 7. The switch classifies each piece of incoming traffic with a priority based on customizable factors. That priority then determines the correct queue for the traffic.

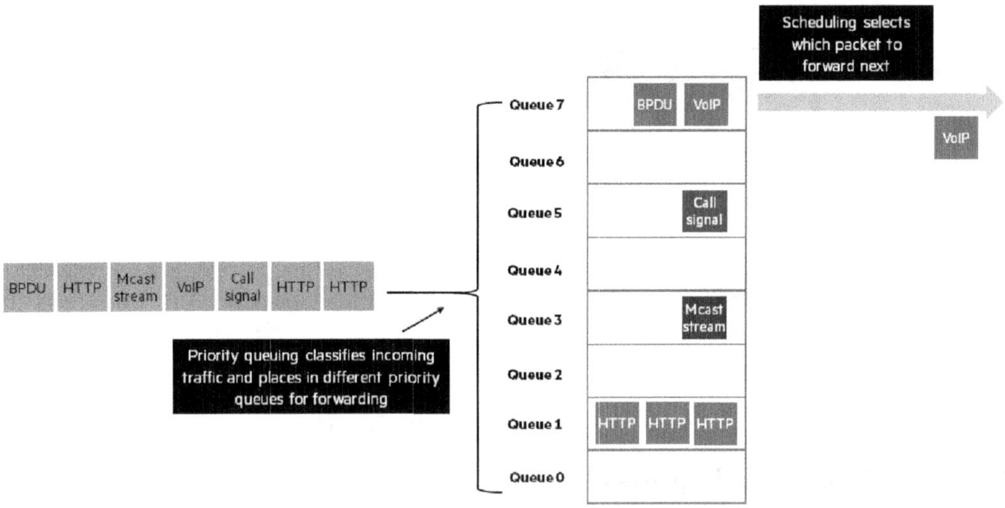

Figure 10-9: AOS-CX queues

FIFO

In the First-In, First-Out (FIFO) queuing method, packets wait in a queue until the switch is ready to process them. In this algorithm, all traffic is regarded equally. If the arrival rate is higher that the processing rate, the queue fills up and any packets that do not fit in the queue are dropped. The oldest entry is processed first; the packets leave the queue in the order they were received.

Figure 10-10: First-In, First-Out

Strict priority queuing (SPQ)

With SPQ, you have two queues in the switch. As the packets enter, they are assigned a priority class. Time-sensitive data is assigned a high-numbered queue and has a high priority and low-numbered queues have a low priority. The packets in the highest priority queue are processed first, and the packets in the lowest priority queue are processed last. This all happens within the AOS-CX switch, and STP provides better QoS than the FIFO queue because higher-priority traffic can reach the destination with less delay. The drawback to this is if there is a continuous stream in a high-priority queue, the packets in the lower-priority queue will never have a chance to be processed. This is called starvation. All AOS-CX switches support SPQ.

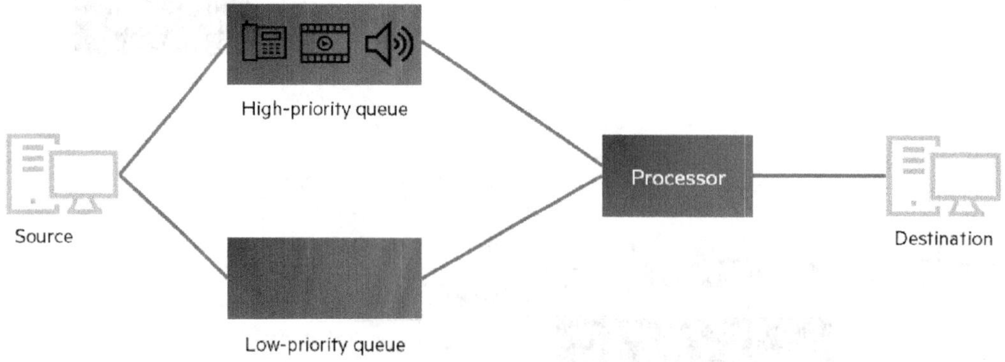

Figure 10-11: Strict priority queuing

Deficit weighted round robin (DWRR)

DWRR uses a deficit counter to allow variable-length packet transmission, preventing queues with larger packets from being unfairly penalized. DWRR allocates available bandwidth among all non-empty queues in relation to the queue weights.

Here is an example of how weighted round robin works. There are three queues: 6, 5, and 1. If you recall, the higher the classification, the higher the priority, and it goes in the higher queue. Each queue has different weights:

- Queue 6 has a weight of 4
- Queue 5 has a weight of 2
- Queue 1 has a weight of 1

The processor goes to queue 6 first because it has the highest priority. It sees the weight of 4 and forwards out all the packets from queue 6. Then it moves to queue 5 and notices it has a weight of 2, so it will only forward two packets. Finally, it moves down to queue 1 and will forward one packet, then back up to queue 6. This is how weighted round robin works.

For deficit weighted round robin, each queue's deficit is set to 0.

1. Queue 6 will be selected for the first request because it has the highest priority (4) and its deficit is 0. Queue 6's deficit will be increased by the total weights of all queues (4 + 2 + 1 = 7).
2. For the second request, the queue with the highest deficit will be selected, which is now queue 5 with a deficit of 7. Queue 5 is selected, and its deficit is increased by the total weights again (7).

3. For the third request, Queue 6's deficit is now 7, queue 5's deficit is 7, and queue 1's deficit is 0. Queue 1 will be selected for the third request, and its deficit will be increased by the total weights (7).

The process continues, with queues being selected based on their deficits and weights until all queues have processed an equal number of requests based on their weights. This way, queues with higher weights will receive more requests over time, maintaining fairness and load balancing in the system.

AOS-CX supports DWRR on all switches.

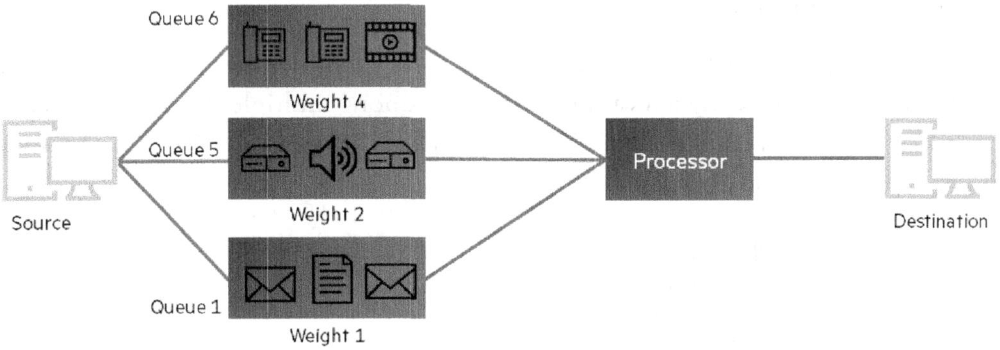

Figure 10-12: Deficit weight round robin

Weighted fair queuing (WFQ)

AOS-CX leverages the WFQ algorithm to provide different weights to each QoS queue, which will support a single or a number of traffic classes in a profile. The profile is called the schedule profile, and by default, all interfaces will have the default schedule profile applied.

Weighted fair queuing allocates available bandwidth among all queues that are not empty, in relation to their queue weights. WFQ is applied in bytes, not packets. It is work conserving, which means that only non-empty queues are counted on each scheduling cycle. The percentage of bandwidth allotted to a non-empty queue is calculated by dividing its weight by the sum of the weights for all non-empty queues. This means that the percentage of bandwidth allotted to a queue can fluctuate, depending on the number of non-empty queues present in each cycle.

Egress queue shaping can be configured using the `max-bandwidth` option to limit the amount of traffic transmitted per output queue at all times, even when there is leftover bandwidth available on the port. The buffer associated with each egress queue stores the excess traffic to smooth the output rate. Sustained rates of traffic above the maximum bandwidth will eventually fill the output queue, causing tail drops.

Use `show interface queues` to determine if any tail-drop errors have occurred.

To remove only egress queue shaping, re-enter the `wfq queue` command without the max-bandwidth parameter. The `no` form of this command removes the weighted fair queuing algorithm from a queue.

The following errors may occur:

- `The max-bandwidth value cannot be greater than 100 percent.` This error occurs when a `max-bandwidth` value greater than 100 percent is configured on a queue.
- `The max-bandwidth cannot be less than <NUM> kbps.` This error occurs when a kbps max-bandwidth value less than the supported minimum kbps shape value is configured on a queue. The supported minimum kbps shape value can be retrieved using the `show capacities` command.

Overall, WFQ is a sophisticated scheduling algorithm that combines fairness with prioritization based on weights. By intelligently allocating bandwidth among multiple flows, WFQ helps optimize network performance, reduce congestion, and ensure a balanced distribution of resources in a network environment. WFQ is supported only on CX 8400 Series switches.

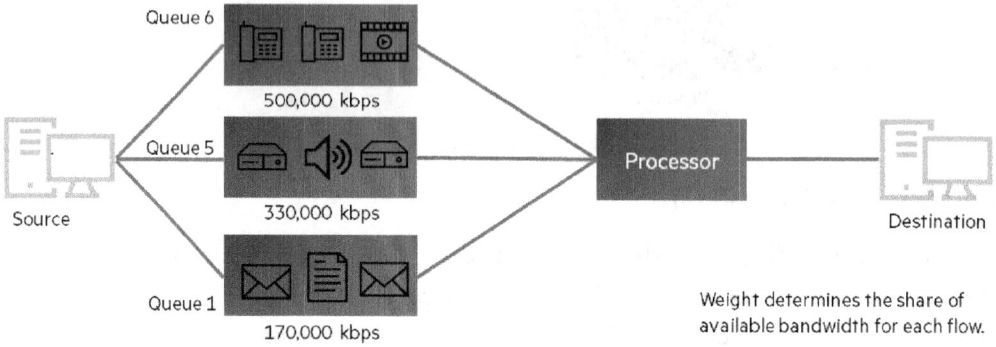

Figure 10-13: Weighted fair queuing

The difference between WRR and WFQ is that WFQ and WRR use different metrics to treat the traffic fairly.

Mixing algorithms

AOS-CX switches also support mixing queue algorithms in the schedule profile.

For example, use SQP on queue 7 to prioritize queue 7 and WRR on queues 6 to 0 to distribute the remaining bandwidth fairly.

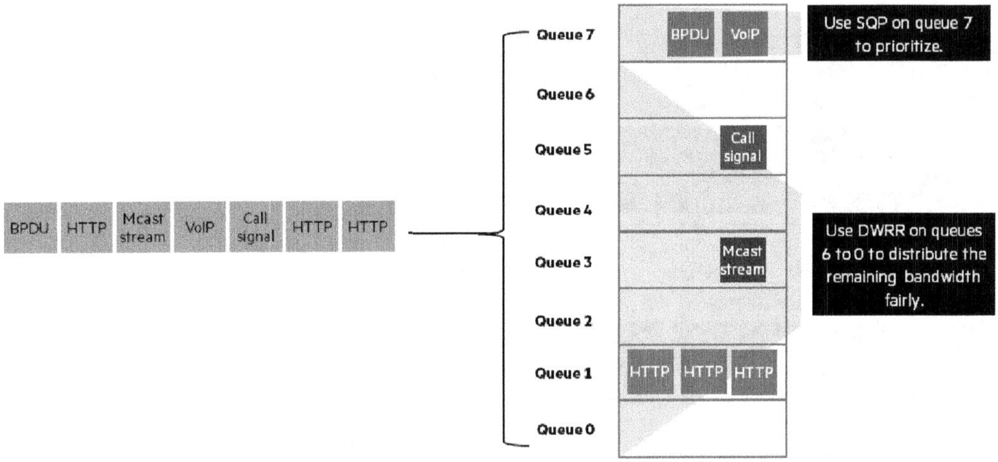

Figure 10-14: Mixing algorithms

QoS processing on AOS-CX switches

Let us explore how AOS-CX switches apply QoS.

When a frame arrives on a port, the switch applies inbound rate limits, if any. It then determines the correct priority or classification for the traffic and assigns it two values: Local Priority (LP) and color. Once a packet is classified using CoS, DSCP, or QoS policy, LP is assigned to each packet. LP helps to determine the correct egress priority queue for the traffic while color determines the traffic's drop eligibility if queues become congested to the point that they must drop traffic.

Next, the switch uses its switching and routing mechanisms to decide the correct egress port for the switch. The switch then queues the traffic at the ingress port for transmission across the switch fabric to that port. Because AOS-CX switches support virtual output queuing (VOQ), an ingress port maintains separate queues for traffic destined to each egress port. In fact, the ingress port has eight priority queues for each egress port. The switch assigns a frame to one of these queues based on its assigned LP and egress port. It then schedules the packet for forwarding across the fabric. There, a similar queuing and scheduling process occurs. Each egress port has eight queues, one for each priority queue, and traffic is assigned to the appropriate one based on its LP. The switch uses a configurable scheduling mechanism to schedule and forward the frame.

You can also apply egress queue shaping in conjunction with strict priority scheduling. This feature sets a rate limit for traffic transmitted per output queue.

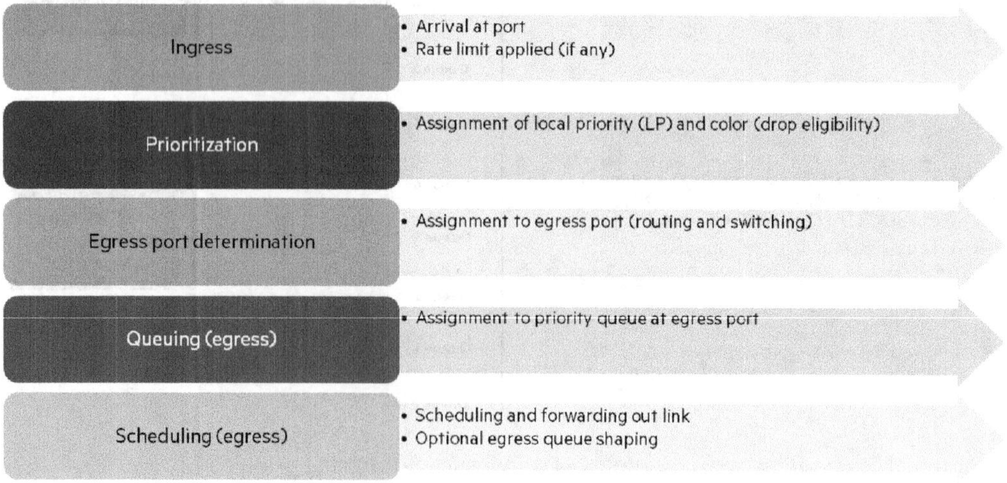

Figure 10-15: QoS processing on AOS-CX switches

Virtual output queues

Virtual output queuing (VOQ) is a technology used in networking switches to enhance packet routing and minimize packet loss. In traditional switch architectures with shared output buffers, multiple input ports contend for access to the same output port, potentially leading to congestion and packet drops. VOQ addresses this issue by assigning a separate virtual output queue to each input port, effectively eliminating contention at the output port.

Here is how VOQ works:

1. **Individual Queues:** In a VOQ-enabled switch, each input port maintains its own virtual output queue. When a packet arrives at an input port, it is placed in the corresponding queue associated with the destination output port.
2. **Decoupling Input and Output:** By decoupling the input and output port queues, VOQ allows packets to be buffered at the input port without contending for the output port. This separation ensures that packets from different input ports do not compete for the same output port at the same time.
3. **Fairness and Prioritization:** VOQ can provide fairness and prioritize certain types of traffic by managing the scheduling and transmission of packets from the virtual output queues. High-priority packets or critical traffic can be given precedence over lower-priority packets.
4. **Reduced Congestion:** Virtual output queues help reduce congestion and minimize packet loss by allowing packets to be buffered and scheduled efficiently. By avoiding head-of-line blocking and contention at the output port, VOQ can improve the overall network performance and quality of service.

5. **Scalability:** VOQ can improve the scalability of switch architectures by providing a more efficient way to handle multiple input ports and diverse traffic patterns. It enables switches to handle higher traffic loads and varying levels of congestion without sacrificing performance.

Overall, VOQ technology plays a crucial role in modern switch designs by optimizing packet handling, reducing contention, and improving the overall efficiency and reliability of network communication. It is particularly beneficial in environments where high performance, low latency, and minimal packet loss are essential requirements. AOS-CX supports VOQ in the CX 6200 Series, CX 6300 Series, CX 6400 Series, and CX 8400 Series.

Rate limiting

Rate limiting is a technique used in networking and computing systems to control the rate of data transmission or processing. It imposes a limit on the number of requests or packets that can be sent within a specific time frame. By implementing rate limiting, organizations can prevent resource exhaustion, maintain system stability, and ensure the fair allocation of resources.

Overview

On AOS-CX switches, port-based rate limiting applies only to flooded traffic—that is, broadcasts, multicasts, and unknown-unicasts. The goal of this rate limiting is to mitigate the effects of broadcast storms or improper network usage while still permitting enough bandwidth to these types of traffic to support their legitimate use. If you want to rate limit unicast traffic, you should set up traffic policing with a classifier-based policy.

The rate limits apply to inbound traffic and are set in terms of packets per second, not bits per second. You set rate limits on physical ports or LAGs. A rate limit on a LAG applies to traffic received across all links in that LAG. You cannot configure an individual rate limit on a physical port.

You can configure different rate limits per port in order to customize the rules as appropriate for the connected devices. For example, multicasts in your network might flow from upstream devices to downstream devices. You could implement a multicast rate limit on the downstream interfaces but not the upstream ones. You could also give the more trusted upstream network a more general broadcast and unknown-unicast rate limit.

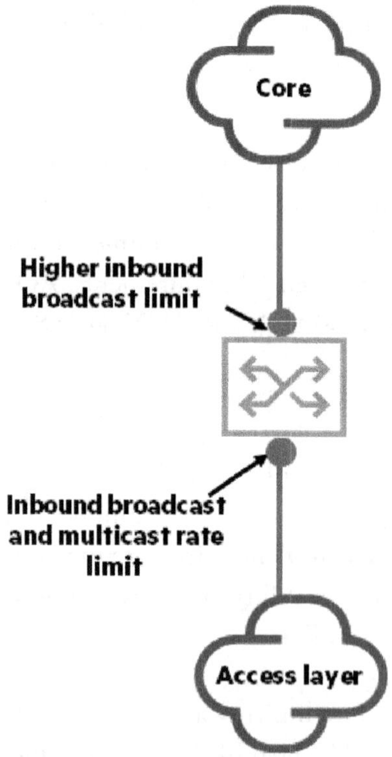

Figure 10-16: Rate limiting

Rate limit configuration

Port rate limiting helps control undesirable traffic. Its purpose is to allow enough unicast, broadcast, multicast, and ICMP traffic for the network to function properly, while preventing flooding and traffic storms.

Some amount of each traffic type is required for normal network operation. For example, broadcast packets may include ARP and DHCP traffic. Video streams and certain types of network protocol packets are multicast traffic. Unknown-unicast packets may be intended for devices whose addresses have temporarily aged out of network forwarding caches. Configuring rate limits can help provide balance between necessary and flooded traffic.

The amount of permitted traffic (the rate limit) can be specified in kilobits per second (kbps), packets per second (pps), or as a percentage of link bandwidth (percent).

Rate limits are enforced separately on each individual member of a LAG, not on the LAG as a whole. When limits are configured in kbps or pps, larger limits may be needed on high-speed ports to allow normal network functionality. A limit of 100

---

pps might be reasonable on a 1 Gigabit port, but it would be too restrictive on a port that is 10 Gigabits or faster. Conversely, limits configured as a percentage of link bandwidth may not be useful on high-speed ports. A limit of 1 percent might be reasonable on a 1 Gigabit port, but too lax on a port that is 10 Gigabits or faster.

A rate limit for multicast traffic will affect both multicast and broadcast traffic, whereas a rate limit for broadcast traffic only impacts broadcast traffic. In the case where both types are applied to the same port, broadcast packets will be limited to the lower of the two configured rate values.

Rate limits configured on LAG interfaces with no assigned or linked-up member ports may be displayed as "not applied." This behavior may also be seen with percentage-based rate limits on regular ports when link up has not yet occurred (that is, when the port speed is not known).

Table 10-1: Rate limit configuration (in Ethernet port or LAG context)

Step	Command	Range/values	Default
Apply a broadcast rate limit	rate-limit broadcast <RATE>	64 – 209090910	No limit
Apply a multicast rate limit	rate-limit multicast <RATE>	64 – 209090910	No limit
Apply an unknown-unicast rate limit	rate-limit unknown-unicast <RATE>	64 – 209090910	No limit

In conclusion, the Quality of Service (QoS) chapter equips participants with essential knowledge and skills to implement QoS mechanisms effectively in network environments. By understanding the key concepts and techniques covered in this chapter, learners will be well-prepared to optimize network performance, prioritize critical applications, and deliver a superior user experience. With a solid grasp of QoS principles, participants are empowered to leverage these tools to minimize latency, ensure reliability, and enhance overall network efficiency.

Learning check

Check your understanding of this material by answering the following questions.

Question 1

Which layer of the network stack is responsible for CoS prioritization?

- O Layer 1
- O Layer 2
- O Layer 3
- O Layer 4

Question 2

Which of the following is a key feature of weighted fair queuing (WFQ)?

- O First-in, First-out (FIFO) scheduling
- O Random packet dropping based on lowest bandwidth
- O Static allocation of bandwidth
- O Fair allocation of bandwidth based on weights

Question 3

What is the definition of jitter in networking and communication systems?

- O The delay variation in packet arrival times.
- O The measure of speed of data transmission.
- O The maximum bandwidth of a network connection.
- O The number of errors in data transmission.

11 Wired network security

Welcome to the "Wired network security" chapter. Network security is a critical aspect of modern-day technology and communication systems. With the increasing reliance on digital networks for various aspects of our personal and professional lives, ensuring the security and integrity of these networks has become more important than ever. Network security helps protect sensitive data, prevent unauthorized access, and safeguard against cyber threats such as hacking, malware, and data breaches. By implementing strong network security measures, organizations and individuals can mitigate risks, maintain privacy, and ensure the smooth and secure operation of their networks. In today's interconnected world, where data is constantly being transmitted and shared, prioritizing network security is essential to safeguarding information and maintaining trust in the digital environment.

In this chapter, you will learn about the basic concepts of network security, as well as the potential risks and threats posed to the organizational data and network infrastructure. You will also learn about port security and how critical it is in preventing unauthorized access to sensitive information. Later in the chapter, common security measures that are used for strengthening network security will be covered.

OBJECTIVES

After completing this chapter, you should be able to:

- ✓ Explain the need for implementing network security measures.
- ✓ Describe common network access security methods.
- ✓ List the various network access control elements.
- ✓ Explain port security and its importance.
- ✓ Describe the 802.1X authentication method.
- ✓ Explain web authentication and its key components.

If you want to know more about network security, please sign up for the *Network Security Fundamentals* course.

Security fundamentals

Network security is the process of implementing hardware and software best practices to protect your network from threats. The CIA triad serves as a foundation for any information security system, including network security. The objective of a secure network is to provide the following:

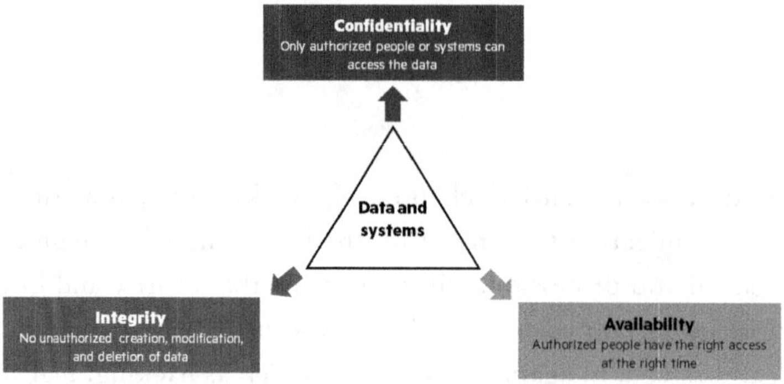

Figure 11-1: The CIA triad

Confidentiality

Only authorized people or systems can access the data. The data must be protected from unauthorized access. This is sometimes referred to as privacy.

Integrity

The data must remain intact when it is stored, processed, and transmitted. Unauthorized creation, modification, and deletion of data must not be allowed. Integrity also refers to data authenticity.

Availability

Systems and data are accessible to authorized users to perform allowed tasks at the right time.

The CIA triad is a well-accepted industry standard that is followed worldwide.

Network access control

Access control is a data security process to manage and restrict access to data and resources. Network access control (NAC) solutions enable you to authorize or prevent users and devices from accessing resources on the network. Here are three main reasons why NAC plays an important role in today's world.

Security: Network access controls protect resources from tampering and theft by malicious actors. NAC solutions ensure that only users and devices with proper permissions can access the network and networked resources. Furthermore, some NAC solutions can identify subjects that may be participating in an attack and quarantine, or block that subject's access pending further investigation. This functionality can prevent the spread of attacks.

Privacy: In today's data-driven world, organizations are managing greater volumes and varieties of data. This includes sensitive and/or confidential data. NAC solutions allow organizations to define who, what, when, and how data can be accessed on the network, to reduce the risk of breach.

Compliance: Regulated organizations often need to comply with data privacy and data protection mandates, such as the General Data Protection Regulation (GDPR), the Health Insurance Portability and Accountability Act (HIPAA), and Sarbanes-Oxley (SOX). NAC solutions can help organizations comply with these mandates by restricting access to data, keeping traffic secure and separate, and providing logging and reporting for audits.

Access control elements

Here are some of the important elements of NAC.

Table 11-1: Elements of NAC

Element	Function	Technologies
Visibility	Knowing who and what is on the network at any given time	Physical or virtual data collectors; active (NMAP, WMI, SNMP, SSH) and passive (SPAN, DHCP, NetFlow/S-Flow/IPFIX) discovery methods; AI/ML assisted device profiling; deep packet inspection
Authentication	Ascertaining with confidence that a user or device is who/what it attests to be	802.1X authentication; EAP-TLS, RADIUS, TACACS+; multifactor authentication; certificates
Policy definition	Defining rules for users and devices regarding resources they can access, and how resources can be accessed	Rule-writing tools, which can include contextual parameters like role, device type, authentication method, device health, traffic patterns, location, and time of day
Authorization	Determining the appropriate rules for the authenticated user or device	N/A
Enforcement	Allowing, denying, or revoking an authenticated user or device's access to a resource based on appropriate policy	Integration and bidirectional communication with firewalls and other security tools

Data encryption

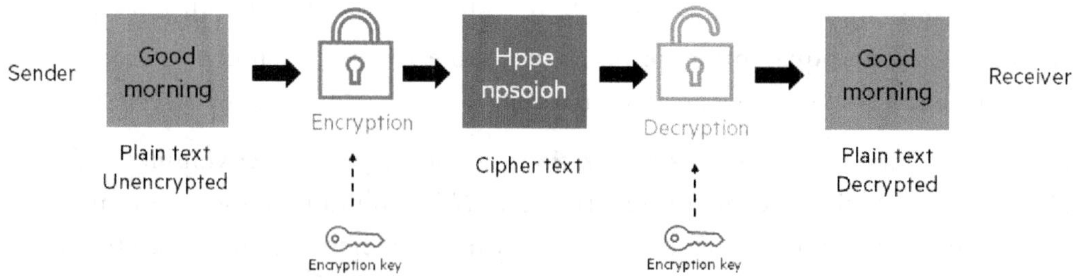

Figure 11-2: Flow depicting the cycle of encryption and decryption of data

Data encryption is an important tool to help protect data. Encryption is the process of converting readable data (known as plain text) into unreadable cipher text. The cipher text can then transit safely across a network. If an unauthorized user eavesdrops on the data, the user will not be able to read the original message without the key. Decryption is the process of converting the cipher text back into plain text with the help of key.

Usually, an algorithm (or a cipher) is used to encrypt and decrypt data. The algorithm contains instructions on how to transform the data. For example, the algorithm might specify substituting one character for another or transposing certain characters. An algorithm works with a key, which is data used by the algorithm to ensure that the same plain text does not always produce the same cipher text. The key is critical because hackers could very well know how a particular encryption algorithm works. It is the key that "locks" the cipher text so that knowledge of the algorithm alone is not enough to allow someone to decrypt the cipher text. Only someone with the correct decryption key can "unlock" the data and return it to plain text.

Physical network security

According to the National Institute of Standards and Technology (NIST), physical safeguards are physical measures, policies, and procedures to protect a covered entity's or business associate's electronic information systems and related buildings and equipment from natural and environmental hazards, and unauthorized intrusion. A key part of network security is ensuring that physical access to your network and its components is monitored closely and regulated. You must take measures to protect your network infrastructure from natural disasters, man-made catastrophes, theft, vandalism, and accidental damage.

Creating a secure environment can help you significantly reduce the risk posed to your network infrastructure. Here are some key design considerations:

- Minimize external access, such as windows and doors. This reduces the surface area prone to risk.
- Increase structural protection. The facility should have high walls and fireproof ceilings.

- Locking devices should be managed responsibly. Periodic maintenance and audit of all the locking devices must be carried out. Whenever a key is compromised, the lock should be changed.
- Maintain an optimal operating temperature and humidity.
- Reduce the visibility of critical facilities. The facility should appear low-key from the outside to ensure unnecessary attention is not received.

Authentication, authorization, and accounting

Authentication, authorization, and accounting (AAA) is a framework used to control and track access in a computer network. AAA provides a uniform framework for implementing network access management. It provides the following security functions:

Authentication: Identifies users and determines whether a user is valid.

Authorization: Grants different users different rights and controls their access to resources and services. For example, a user who has been successfully authenticated can be granted read/write permissions to the files on the device.

Accounting: Records all user network service usage information, including the service type, start time, and traffic. The accounting function not only provides the information required for charging, but also allows for network security surveillance.

Wired network authentication

In the context of networking authentication, particularly involving protocols like 802.1X, it is important to understand the roles of different components involved in the authentication process. Here is a detailed explanation of the differences between a supplicant, Network Access Device (NAD), Network Access Server (NAS), and authentication server.

Supplicant

Definition: The supplicant is the client device or the software running on the client device that requests access to the network.

Role: It initiates the authentication process by providing credentials (e.g., username/password, certificate) to gain network access.

Functionality: The supplicant sends EAP (Extensible Authentication Protocol) messages to the authenticator (NAD/NAS) to begin the authentication dialogue. It interacts with the authenticator to forward its credentials to the authentication server.

Examples: Laptops, smartphones, tablets, or any device with 802.1X supplicant software.

Network Access Device (NAD)

Definition: A NAD is any network device that provides connectivity to the network for end-user devices.

Role: Acts as an entry point to the network, controlling access based on authentication results.

Functionality: It interacts with the authentication server to grant or deny access requests. In the context of 802.1X, it forwards EAP messages between the supplicant and the authentication server.

Examples: Switches, wireless access points.

NAS (Network Access Server)

Definition: A Network Access Server is a specialized device that provides centralized authentication, authorization, and accounting (AAA) services for users who want to connect to a network.

Role: The NAS acts as an intermediary between the end-user devices and the authentication servers.

Functionality: The NAS is an intermediary between the end-user or NAD devices and the authentication servers. It often communicates with authentication servers (like RADIUS or TACACS+ servers) to verify user credentials and determine what level of access should be granted.

Examples: HPE Aruba Networking AOS-10 Wireless gateways acting as RADIUS proxy.

As AOS-CX switches often perform both roles as NAD and NAS, it is common to refer to the switch as either NAD or NAS during the authentication process. NAD and Nas could also be referred to as the authenticator device.

Authentication Server

Definition: The authentication server is a centralized server responsible for validating the credentials provided by the supplicant.

Role: It determines whether the supplicant's credentials are valid and, optionally, what level of access should be granted.

Functionality: It receives EAP messages from the authenticator, validates the credentials against its database, and sends an EAP success or failure message back to the authenticator (NAD/NAS).

Examples: RADIUS server, TACACS+ server.

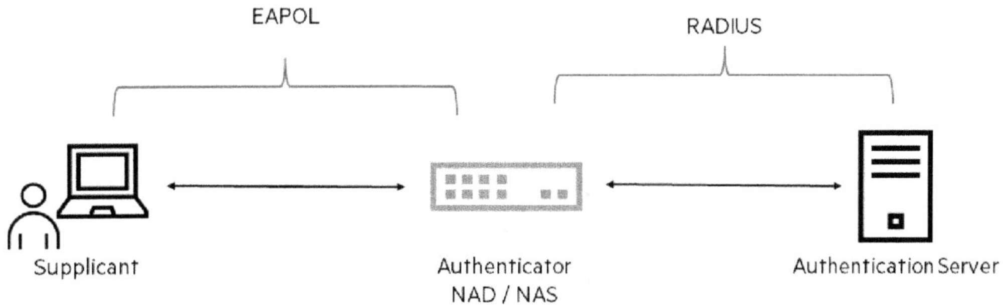

Figure 11-3: Roles of wired network authentication

In summary, while the supplicant, NAD, NAS, and authentication server all play roles in network authentication, their functions and responsibilities differ significantly. The supplicant is the client seeking access, the NAD controls the entry point, the NAS (in specific scenarios) manages remote access, and the authentication server validates credentials.

Local and remote authentication

AAA usually uses a client/server model. The client runs on the Network Access Server (NAS), and the server maintains user information centrally. In an AAA network, a NAS is a server for users but a client for the AAA servers. Authentication of access to HPE Aruba Networking products can be done via one of two methods: local or remote. Remote authentication is sometimes referred to as centralized authentication.

Local authentication

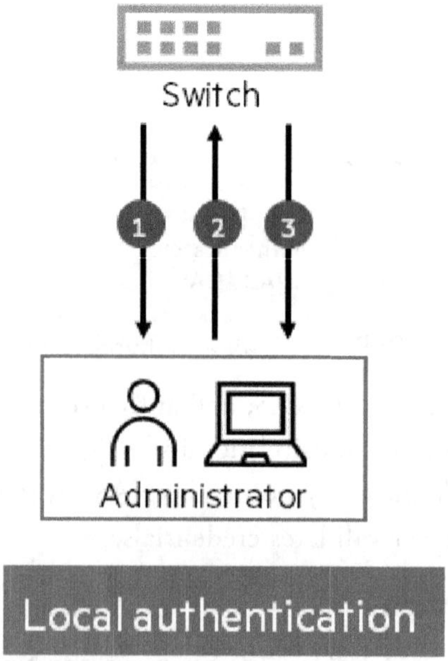

Figure 11-4: Topography depicting the steps involved in local authentication

In local authentication, if a network administrator attempts to access a switch and credentials are configured on the switch, the following occurs:

1. The switch prompts the administrator to enter credentials.
2. The administrator enters the credentials.
3. The switch grants access if the administrator enters the correct credentials; the switch blocks access if the credentials are incorrect.

Remote authentication

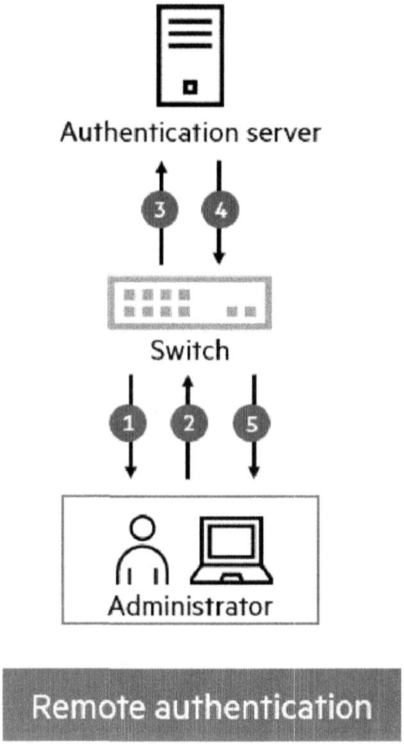

Figure 11-5: Topography depicting the steps involved in remote authentication

If a network administrator attempts to access a switch and remote authentication is configured, the following occurs:

1. The switch prompts the administrator to enter the credentials.
2. The administrator enters the credentials.
3. The switch forwards the login credentials to the RADIUS or TACACS+ server.
4. The server validates the login credentials against its database and notifies the switch if they match. It also notifies the switch what privilege level the administrator should be given.
5. The switch grants the appropriate access level.

Authentication is critical for controlling who accesses the networking devices and the tasks they can perform. The following two sections briefly introduce these topics as they relate to the two series of HPE Aruba Networking products.

There are two common protocols networking devices might employ when interacting with an AAA server:

- Remote Authentication Dial-In User Service (RADIUS)
- TACACS+

RADIUS authentication

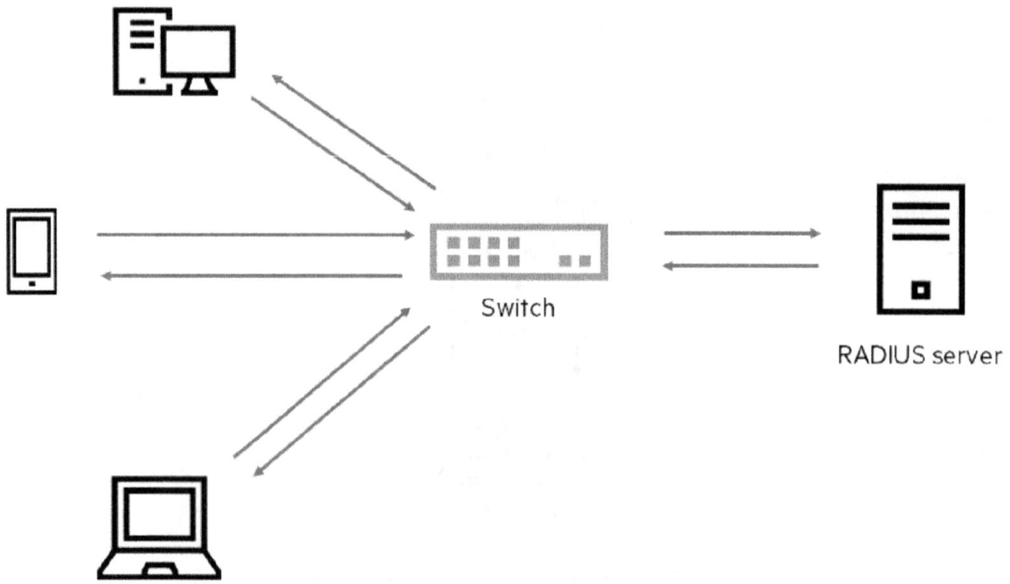

Figure 11-6: Elements involved in RADIUS authentication

RADIUS is a networking protocol used to manage and secure remote access to networks. It provides a centralized AAA system for network devices, such as routers, switches, and wireless access points. The importance of RADIUS in network security stems from its ability to enhance authentication, control access, and monitor user activities.

Here are some key points highlighting the importance of RADIUS:

- **Centralized authentication:** RADIUS allows organizations to manage user authentication centrally for remote access to the network. This ensures that only authorized users can connect to the network, reducing the risk of unauthorized access and potential security breaches.
- **Access control:** RADIUS provides granular control over user access privileges. It enables organizations to define and enforce access policies based on user roles, groups, or other criteria. This helps prevent unauthorized users from accessing sensitive resources or data.
- **Stronger security:** RADIUS supports a variety of authentication methods, including strong cryptographic protocols like Extensible Authentication Protocol (EAP). By using RADIUS, organizations can implement stronger security measures, such as two-factor authentication, to further protect network access.

- **Accountability and auditing:** RADIUS enables organizations to track and log user activities, providing an audit trail for monitoring and compliance purposes. This helps in identifying any suspicious or unauthorized activities and aids in forensic investigations if necessary.
- **Scalability:** RADIUS is highly scalable, allowing organizations to add or remove users, devices, and access policies easily as their network grows or changes. It provides a flexible and efficient solution for managing large numbers of users across multiple network devices.

In summary, RADIUS is a critical component of network security as it provides centralized control, robust authentication, access control, and auditing capabilities. By implementing RADIUS, organizations can enhance their network security, protect sensitive resources, and ensure compliance with industry regulations.

RADIUS uses UDP to share information between the NAS and the security server. The NAS and RADIUS server are configured with a common key that they use to create a message digest of some of the information they exchange. This rudimentary encryption method is somewhat inefficient, making the exchanged information susceptible to eavesdropping. Another problem with RADIUS is that many additions have been made to the protocol by many different vendors, which often creates incompatibility problems. RADIUS uses UDP, and its packet format and message transfer mechanism are based on UDP. It uses UDP port 1812 for authentication and port 1813 for accounting. RADIUS has an important advantage over the other two protocols. It has a more robust accounting system and, therefore, is very common for user connections, such as remote access VPNs, 802.1X, and dial-up services.

TACACS+

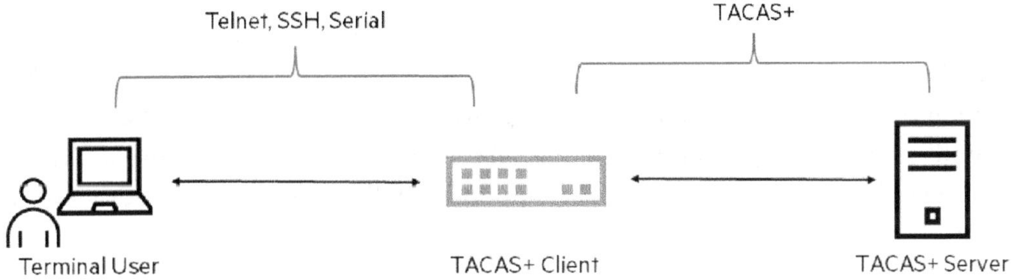

Figure 11-7: TACACS+ Terminal user communication to TACAS+ Server

TACACS+ (Terminal Access Controller Access Control System) is a protocol used to manage network access control and authentication. TACACS+ is similar to RADIUS, but it is used for management protocols. Originally developed by Cisco, TACACS+ is a protocol used for centralized authentication, authorization, and accounting (AAA) for users who access a network. It is particularly used for administrative access to network devices and provides a way to manage network access control.The importance of TACACS+ lies in its ability to provide a secure and efficient method for controlling access to network devices and resources. By using TACACS+, organizations can enforce user authentication, assign specific levels of access privileges, and track user activity for auditing and compliance purposes. Some key benefits and importance of TACACS+ include:

- **Centralized Control:** TACACS+ allows organizations to centrally manage user authentication and authorization policies for network devices, ensuring consistent access control across the network.
- **Enhanced Security:** TACACS+ provides a secure method for authenticating users and controlling access to sensitive network resources, helping to prevent unauthorized access and potential security breaches.
- **Accountability:** TACACS+ enables organizations to track and log user activities on network devices, providing a detailed audit trail for monitoring and compliance purposes.
- **Scalability:** TACACS+ supports the scalability of network access control by allowing organizations to easily add or remove users, devices, and access policies as needed.

Overall, TACACS+ plays a critical role in enhancing network security, improving operational efficiency, and ensuring regulatory compliance for organizations of all sizes.

Port access

Port access is a critical component of network security, aimed at safeguarding network resources and preventing unauthorized access to sensitive information. In essence, port access involves controlling and monitoring the physical and logical access to network ports on switches and routers. LANs are often deployed in a way that allows unauthorized clients to attach to network devices or allows unauthorized users to get access to unattended clients on a network. 802.1X and MAC authentication simplify security management by providing access control plus the ability to control user profiles, allowing a given user entering valid user credentials access from multiple points within the network. AOS-CX access switches support the following port access authentication types:

- 802.1X authentication
- MAC authentication
- Captive Portal

Port based authentication—802.1X

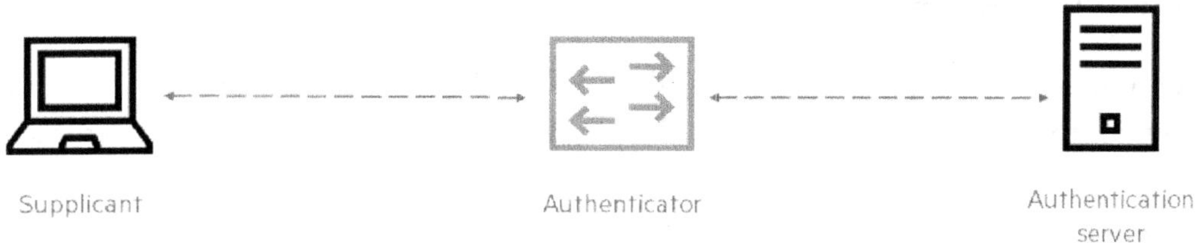

Figure 11-8: Information flow between supplicant, authenticatior, and the authentication server

The Institute of Electrical and Electronics Engineers (IEEE) developed the 802.1X standard for port-based authentication. The 802.1X standard provides administrators with an authentication mechanism for devices trying to access a LAN or WLAN. 802.1X defines the encapsulation of the EAP over IEEE 802, which is known as EAP over LAN (EAPOL). Using this option, the port processes all IP traffic as if it comes from the same client. The key components of the 802.1X authentication process are: Supplicant, Authenticator, and Authentication server.

To understand these components, let us consider an analogy in which a person named Bob is attempting to get into an exclusive, very important person (VIP) event in a secure building.

Supplicant

The device that tries to access the LAN. The supplicant is the 802.1X component that runs on the endpoint attempting to connect to the network. In our analogy, Bob, the person attempting to get into the event, is the supplicant. He is the one who wants to gain access to something. In the real world, 802.1X supplicants are software that runs on client devices. For example, Windows and MAC operating systems have native 802.1X supplicants, as do smartphone operating systems such as Android. Admins can also install third-party supplicant software on endpoints.

Authenticator

A network device, such as an Ethernet switch, that authenticates the supplicant. The authenticator controls network access. It enforces decisions made by the authentication server, either permitting or denying access. In our analogy, the authenticator is the security guard at the building in which the event is being held. The security guard is responsible for guarding the building and only allowing authorized people in. Similarly, in 802.1X, the authenticator stands at the internal network perimeter and either grants or denies access to the network. In HPE networks, the authenticator could be a switch, an AP, or an HPE Aruba Networking gateway.

Authentication server

A host running software supporting the RADIUS and EAP protocols that provides an authentication service to the authenticator. The authentication server directs the authenticator and ultimately takes responsibility for access decisions. You can think of the authentication server as the event organizer, who has the VIP list and tells the security guard who to let in. In 802.1X, the authentication server is a RADIUS server. To make its access decisions, the RADIUS server checks users' credentials, but also considers other factors. Is the user connecting at the right time or location? An authentication server may, for example, deny users with legitimate credentials if they are trying to connect outside of regular work hours. The decisions all depend on the authentication server's policies. It is recommended that you use HPE Aruba Networking ClearPass Policy Manager Platform.

MAC authentication

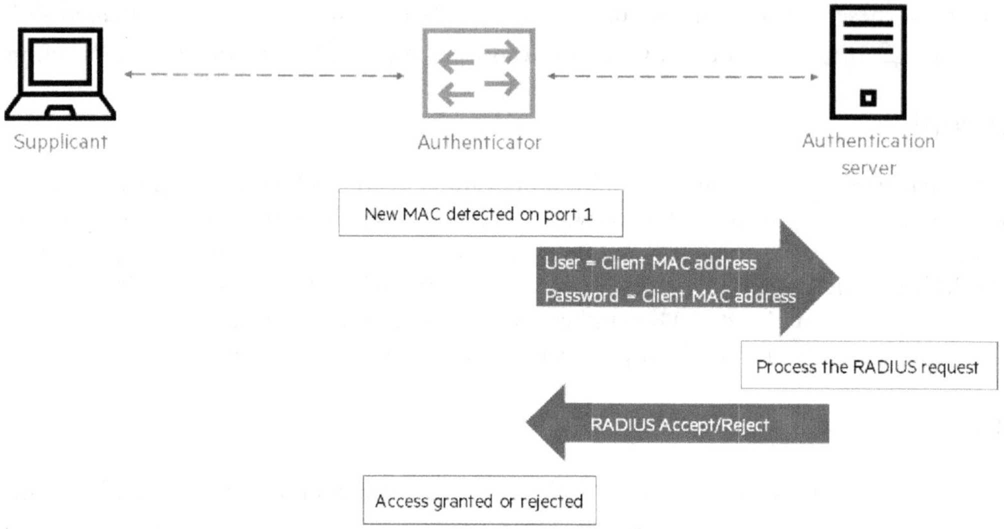

Figure 11-9: Various steps involved in MAC authentication between supplicant, authenticator, and the authentication server

MAC authentication is a security method used in computer networks to control access to the network based on the physical address of a device's network interface card (NIC). Each NIC has a unique MAC address assigned by the manufacturer, making it a useful identifier for network security purposes. MAC authentication is important in network security because it helps to ensure that only the MAC address list in the authentication server can connect to the network. By configuring network devices to only allow connections from specific MAC addresses, organizations can prevent unauthorized access, reduce the risk of network intrusions, and protect sensitive data from being accessed by unauthorized users. Additionally, MAC authentication can also be used to track and monitor network activity, troubleshoot network issues, and enforce security policies within the network environment. MAC authentication is not a recommended form of security.

In highly secured networks MAC authentication is not a best practice for use because the process is easily spoofed and can allow intruders easy access to the network.

Concurrent authentication

MAC authentication with 802.1X

You can configure MAC authentication with 802.1X on a port. When both options are configured, the combined authentication method will have the following features:

- The default precedence order is 802.1X authentication, MAC authentication, device profile.
- When a higher precedence authentication method fails, the profile of the next highest precedence authentication method (that is successful) is applied.
- If all authentication methods fail, the reject role/critical role will be applied based on the 802.1X failure reason, and authentication attempts continue with 802.1X if enabled.
- Upon successful authentication, the client profile is applied based on the authentication precedence configured for the port.
- Concurrent onboarding causes all the authentication methods—such as 802.1X and MAC—to start concurrently, enabling faster onboarding. Port access concurrent onboarding enables faster client onboarding. Concurrent authentication enables all methods to start concurrently for a faster onboarding process.

Example

Figure 11-10: Topography showing PC and IP phone connected to a switch

In many campus deployments, just one Ethernet cable is laid out for each user desk, where an IP phone and a PC/laptop must be connected to the network. IP phones usually have two Ethernet ports, one connected to the switch and another available to be connected to another device, such as a PC. As many IP phones do not have support for 802.1X, the switch will authenticate the phone through its MAC address, while the PC will authenticate normally using 802.1X. The switch keeps track of each MAC address present in the port and its authentication status.

Port security

What is port security?

Port access is a critical component of network security, aimed at safeguarding network resources and preventing unauthorized access to sensitive information. In essence, port security involves controlling and monitoring the physical and logical access to network ports on switches and routers.

Why is port security important?

Port security enables you to configure each switch port with a unique list of the MAC addresses of devices that are authorized to access the network through that port. This security enables individual ports to detect, prevent, and log attempts by unauthorized devices to communicate through the switch.

- Default port security operation: The default port security setting for each port is the dynamic mode, in which the switch learns addresses from inbound traffic from any connected device.
- MAC Lockdown: Also known as static addressing, MAC Lockdown is used to prevent station movement and MAC address hijacking, by allowing a given MAC address to use only an assigned port on the switch. MAC Lockdown also restricts the client device to a specific VLAN.
- MAC Lockout: This enables blocking a specific MAC address so that the switch drops all traffic to or from the specified address.

Port security does not prevent intruders from receiving broadcast and multicast traffic. MAC Lockdown has a higher priority over port security. We have port security and port access that complement each other.

Port security general options

Let us look at some of the general options available to configure port security. On a per-port basis, you can configure security measures to block unauthorized devices and to send notice of security violations. Once port security is configured, you can then monitor the network for security violations through one or more of the following means:

- Alert flags captured by network management tools
- Alert Log entries in the WebAgent
- Event Log entries in the console interface

For any port, you can configure the following options:

- **Action:** Used when a port detects an intruder. Specifies whether to send an SNMP trap to a network management station and whether to disable the port.
- **Address limit:** Sets the number of authorized MAC addresses allowed on the port.
- **Static:** Enables you to set a fixed limit on the number of MAC addresses authorized for the port and to specify some or all the authorized addresses. (If you specify only some of the authorized addresses, the port learns the remaining authorized addresses from the traffic it receives from connected devices.)
- **Configured:** Requires that you specify all MAC addresses authorized for the port. The port is not allowed to learn addresses from inbound traffic.
- **Authorized (MAC) addresses:** Specify up to eight devices (MAC addresses) that are allowed to send inbound traffic through the port. This feature closes the port to inbound traffic from any unauthorized devices that are connected to the port. It also provides the option for sending an SNMP trap notifying of an attempted security violation to a network management station and, optionally, disables the port.
- **Port access:** Allows only the MAC address of a device authenticated through the switch's 802.1X port-based access control.

Dynamic Segmentation

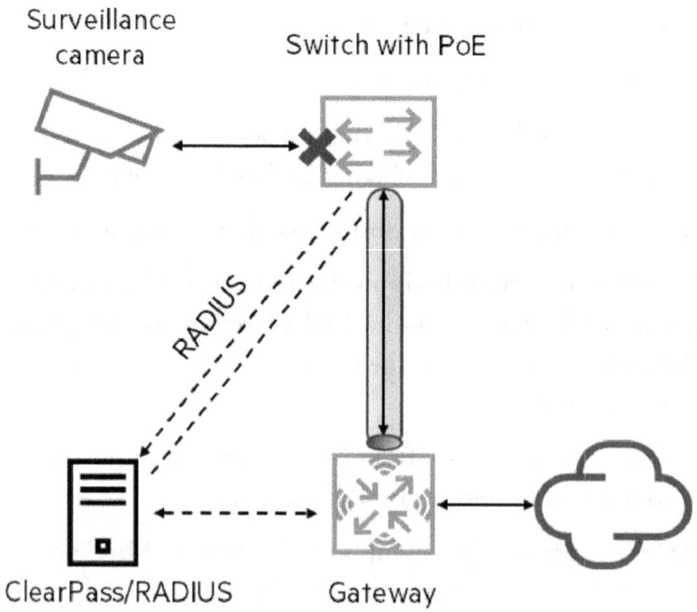

Figure 11-11: The flow of information between various elements when dynamic segmentation is enabled

All modern desktops, laptops, tablets, and smart phones support robust security mechanisms like 802.1X/EAP. But many devices lack these security mechanisms. This includes specialty devices and some devices that rely on PoE, such as security cameras, payment card readers, IoT, and medical devices. This lack of endpoint security can pose a serious risk to your infrastructure.

A switch with Dynamic Segmentation improves your security stance while easing administration. It authenticates these devices to a RADIUS server, and tunnels the traffic to an HPE Aruba Networking gateway, which centralizes formerly disparate access control lists (ACLs) and firewall rules into a single, intuitive policy system. You can continue to use these devices without compromising network security.

Dynamic Segmentation—operation overview

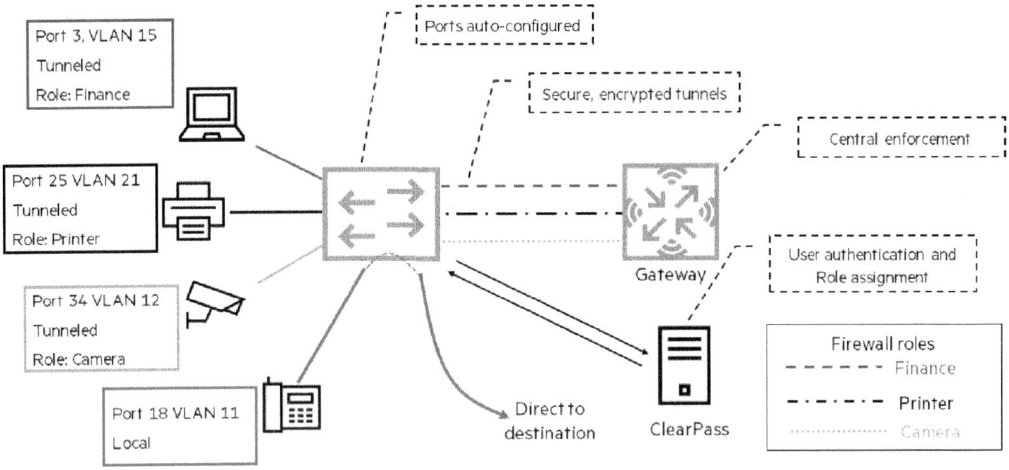

Figure 11-12: Operational overview of Dynamic Segmentation

ClearPass profiles each device, deems appropriate endpoints as "acceptable," and sends accept messages to the controller. The messages include tunnel information, assigned VLAN, and secondary role. In the figure, a user was profiled into the Finance role, and so their port is configured for VLAN 15, and their traffic is securely tunneled to the controller. The controller applies centralized policy enforcement to limit finance traffic as appropriate. Printers and cameras are similarly treated, as shown in **Figure 11-12**.

ClearPass can also indicate if device traffic should be locally forwarded by the switch, directly toward its ultimate destination, without tunneling to the controller. You sacrifice some of that centralized policy enforcement firewall (PEF) enforcement, but you accommodate devices that are uber-sensitive to delay and jitter. You just do not get all the benefits of this "tunneled-node" solution.

Captive portal

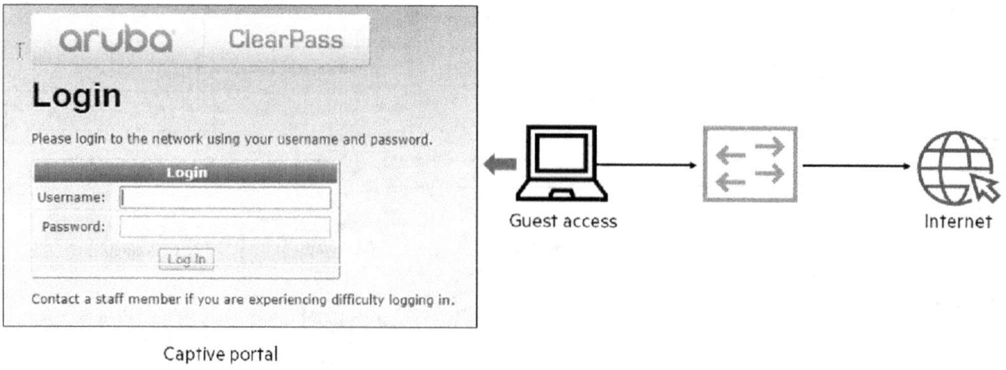

Figure 11-13: An example of guest access through a captive portal

Captive portal, also known as web authentication, is a web-based solution that provides an alternative authentication methods for devices other than 802.1X or MAC Authentication. For example, guests in a hotel or meeting room do not have a corporate credential that allows them to authenticate using 802.1X, and the IT department has no way of knowing the MAC address of each guest device to authorize its access. Therefore, captive portal authentication uses the client browser, redirecting the guests to an authentication page. Some solutions, such as ClearPass or HPE Aruba Networking Central, can authenticate guests who already have an account but also provide a self-registration option to new guests, providing an easy and intuitive way for guests to access the network.

Captive portal provides clients with internet access based on http or https redirection. The client first gets authenticated using MAC or 802.1X authentication, which results in application of the captive portal profile. The HTTP or HTTPS request is then redirected to the captive portal server for user registration. Finally, with Change of Authorization (CoA) using the Port Bounce VSA or Disconnect Message from the RADIUS server, MAC or 802.1X authentication occurs, providing the authenticated client with internet access. Captive portal is supported only for the clients getting authenticated through RADIUS servers.

Captive portal—use case and certificate

Use case

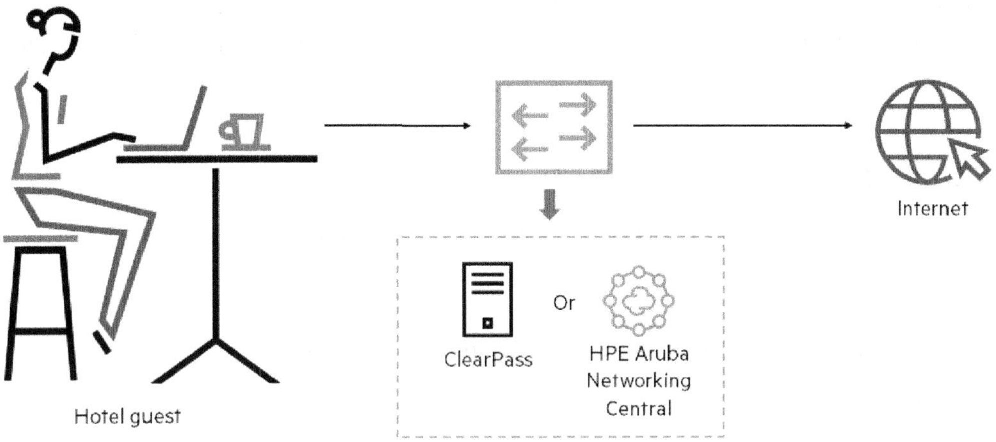

Figure 11-14: Placement of ClearPass or HPE Aruba Networking Central in captive portal authentication process

A common use case for captive portal authentication is authenticating guests in a hotel network. It simplifies the user experience since there's no need to configure their devices or install apps or programs. In a hotel network, when guests connect to the WiFi or an Ethernet port available in the room, their browser session will be automatically redirected to an authentication page. The guest may authenticate using the credentials provided by the front desk or register a new user. Once authenticated, the guest will access the internet and specific local resources defined by the network admin.

Certificate

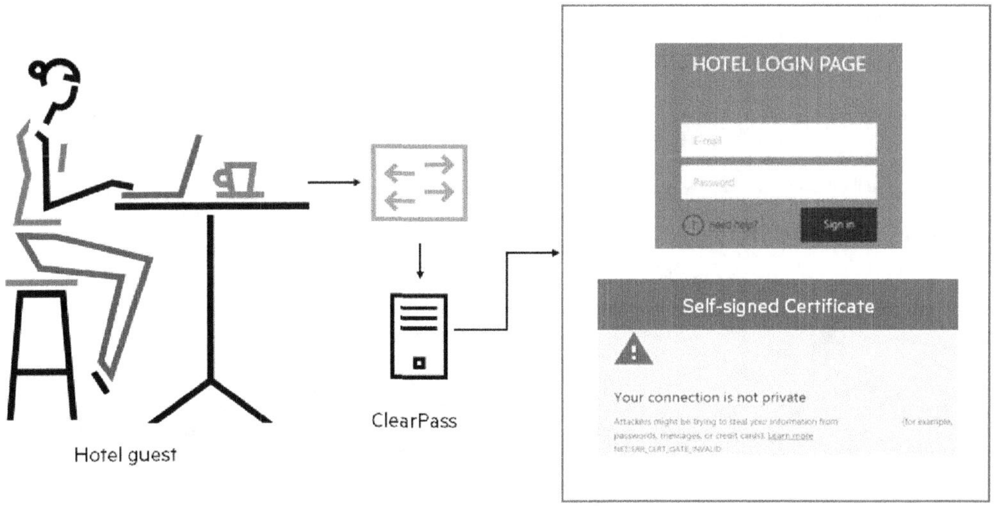

Figure 11-15: Sample web interface of a captive portal showing warning message

A best practice is to use HTTPS for captive portals, as it encrypts the traffic that contains the user credentials. The use of a self-signed or enterprise CA-signed certificate is not recommended, as guests do not have the signing CA in their trust list. A "connection not private" warning message will be displayed, which will cause many users to abort the process. Therefore, HPE recommends that public-signed CAs should be used for captive portal authentication.

Learning check

Check your understanding of this material by answering the following questions.

Question 1

What are the three basic objectives of network security? (Select three)

- ☐ Availability
- ☐ Confidentiality
- ☐ Integrity
- ☐ Transmissibility

Question 2

Which component of the AAA framework involves checking what a user is allowed to do?

- ○ Authentication
- ○ Authorization
- ○ Accounting
- ○ Availability

12 Secure management and

Welcome to the "Secure management and maintenance" chapter. Secure management and maintenance for switches is crucial in safeguarding the integrity and reliability of network operations. By implementing robust security measures and proactive maintenance routines, organizations can mitigate potential threats and vulnerabilities that may compromise switch functionality. This chapter will explore key strategies and best practices for ensuring the secure management and maintenance of switches to protect critical network infrastructure effectively.

OBJECTIVES

After completing this chapter, you should be able to:

- ✓ Describe the concept of Out-of-Band Management (OOBM).
- ✓ Explain secure management protocols such as Secure Shell (SSH) and Hypertext Transfer Protocol Secure (HTTPS), which are used to establish encrypted communication channels.
- ✓ Use Simple Network Management Protocol (SNMP) in monitoring and managing switches efficiently.
- ✓ Apply skills in managing switch OS images securely.
- ✓ Implement comprehensive security measures and maintenance practices.

Secure management

Secure management for switches is a critical aspect of network infrastructure security. It involves key elements such as using an OOBM port and creating a management VRF to isolate management traffic. Access to management functions should be restricted to authorized users through secure protocols like SSH, HTTPS, and the Web Interface, with robust AAA measures in place. Role Based Access Control, RADIUS authentication, and SNMP monitoring are employed for precise user permissions and activity tracking, while secure configuration file management practices protect switch configurations. You will learn more about these key elements in this chapter.

In-band and out-of-band management

Management communications with a managed switch or network device can be either of the following:

- In band
- Out of band

Figure 12-1: An HPE Aruba Networking switch with in-band and out-of-band management ports

Let us look at some of the features of both types of management types.

In-band management

- In-band management involves using the same network infrastructure that is used for regular data traffic to manage and monitor network devices.
- Management traffic shares the same network path as user data, which can sometimes lead to network congestion and potential security risks.
- If the network is down or experiencing issues, in-band management may become unavailable or less reliable.
- In-band management is typically easier to implement and configure compared to out-of-band management.

Out-of-band management

- Out-of-band management involves using a separate, dedicated network for managing and monitoring network devices.
- This dedicated network provides a separate path for management traffic, ensuring that it is not affected by congestion or issues on the main network.

- Out-of-band management is more secure as it is isolated from the regular data traffic, reducing the risk of unauthorized access or attacks.
- Even if the main network is down, out-of-band management can still be used to access and troubleshoot network devices.

In conclusion, in-band management is simpler to implement but may be less reliable and secure compared to out-of-band management. Out-of-band management provides a more secure and reliable way to manage network devices, especially in critical environments where uptime and security are paramount.

Out-of-band management port

Starting at the CX 6200 Series and up, AOS-CX switch families include an OOBM port, used exclusively to monitor and manage the switch. AOS-CX has a separate management VRF, and because of this there is complete isolation between the management and data planes. Data traffic can never see or use this port. Due to this complete isolation, management traffic does not use any bandwidth in the data plane. This is commonly referred as Out-of-Band Management, or OOBM.

Figure 12-2: AOS-CX switch with OOBM port

You do have the option of remotely connecting to and managing the switch via a normal Ethernet port in the data plane. This is referred to as In-Band Management, and it is often used in Campus networks as the cost to build an entirely separate network that spans multiple IDFs, MDFs, and buildings will increase the implementation cost. In-band management is usually protected by using a separate subnet for management that can be protected with ACLs, Firewalls, and COPP policies.

Management VRF

You learned about Virtual Routing and Forwarding (VRF) in a previous chapter of this guide. VRF creates separate virtual routers inside a physical router, with separate routing tables. AOS-CX devices have a default VRF for the data plane, and a separate management VRF for the management port to handle OOBM traffic. The AOS-CX management interface is enabled by default.

The CX 4100i Series, CX 6000 Series, and CX 6100 Series do not have an OOBM port, therefore they do not have the mgmt VRF.

It is configured to receive its IP parameters using DHCP. Use the `show interface management` command to verify the state of this interface.

If preferred, you can manually configure IP parameters on the management interface when not using DHCP, as follows.

```
Switch(config)# interface mgmt
Switch(config-if-mgmt)# ip static <IP-ADDRESS/MASK>
Switch(config-if-mgmt)# default-gateway <DEFAULT-GATEWAY-IP>
Switch(config-if-mgmt)# nameserver <DNS-PRIMARY-SERVER>
```

The following is a sample output.

```
Switch# show interface mgmt
Address Mode                 : static
Admin State                  : up
Mac Address                  : 88:3a:30:97:c7:01
IPv4 address/subnet-mask     : 10.251.1.4/24
Default gateway IPv4         : 10.251.1.254
Primary Nameserver           : 10.253.1.21
```

Ping and traceroute in management VRF

By default, the AOS-CX switch ping and traceroute commands are generated for the data plane's default VRF. To verify connectivity on the management VRF, you must use special ping and traceroute syntax to specify the management VRF, as shown in the following example.

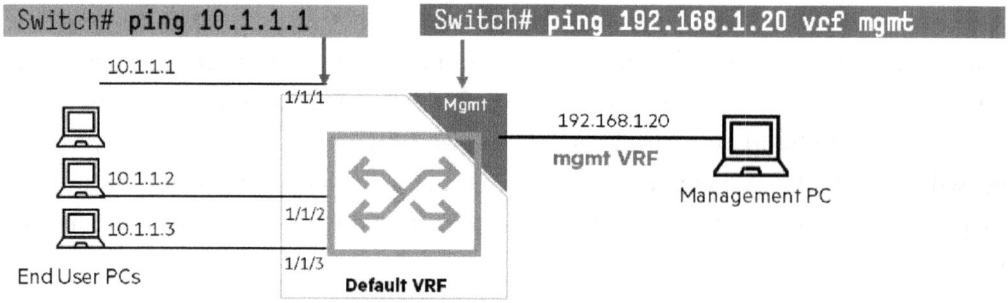

Figure 12-3: Ping and traceroute in management VRF

SSH for AOS-CX

SSH provides secure communications between the switch and your management PC. The following list describes the default setting on various AOS-CX switches:

- For the CX 4100i Series, CX 6000 Series, and CX 6100 Series, SSH is enabled by default on the default VRF.
- For the CX 6200 Series, CX 6300 Series, and CX 6400 Series series SSH is enabled by default on the management and default VRF.
- For the CX 8xxx Series, CX 9300 Series, and CX 10000 Series, SSH is enabled by default on the management VRF.

Use the following syntax to enable SSH for VRF management.

```
Switch(config)# ssh server vrf mgmt
```

Use the `show ssh server vrf mgmt` command to check SSH status on the management interface. Once SSH is enabled in the management plane, many administrators prefer to disable SSH in the data plane. Although SSH is secure and requires proper authentication credentials, disabling SSH in the data plane ensures that SSH connectivity is not possible for end users and potential bad actors.

HTTPS for AOS-CX

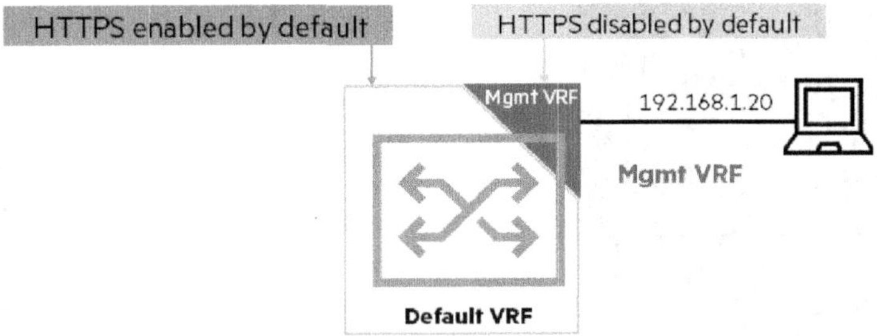

Figure 12-4: HTTPS enabled/disabled on different planes

The following list describes the default settings:

- The CX 6200 Series, CX 6300 Series, and CX 6400 Series switches have the HTTPS server enabled on the management and default VRF.
- The CX 4100i Series, CX 6000 Series, and CX 6100 Series switches have the HTTPS server enabled on the default VRF.
- The CX 8xxx Series, CX 9300 Series, and CX 10000 Series switches have the HTTPS server enabled by default on the management VRF.

The objective for HTTPS for AOS-CX is to have a secure protocol for switch GUI access.

Treat HTTPS like SSH – enable it in the management plane using the following syntax. Then disable it in the data plane's default VRF to maximize security.

```
Switch(config)# https-server vrf mgmt
```

Enter the `show https-server` command to verify the HTTPS status.

To disable the HTTPS service in the VRF default, enter the following command:

```
Switch(config)# no https-server vrf default
```

Web interface

AOS-CX switches integrate a GUI web interface for device monitoring and management. To use this interface, open a web browser and enter the switch IP address: https://switch-ip-address. By default, the AOS-CX switch will have self-signed certificates. It will prompt a certificate error. It is recommended to install a signed certificate when managing AOS-CX switches with HTTPS. The login screen is as follows:

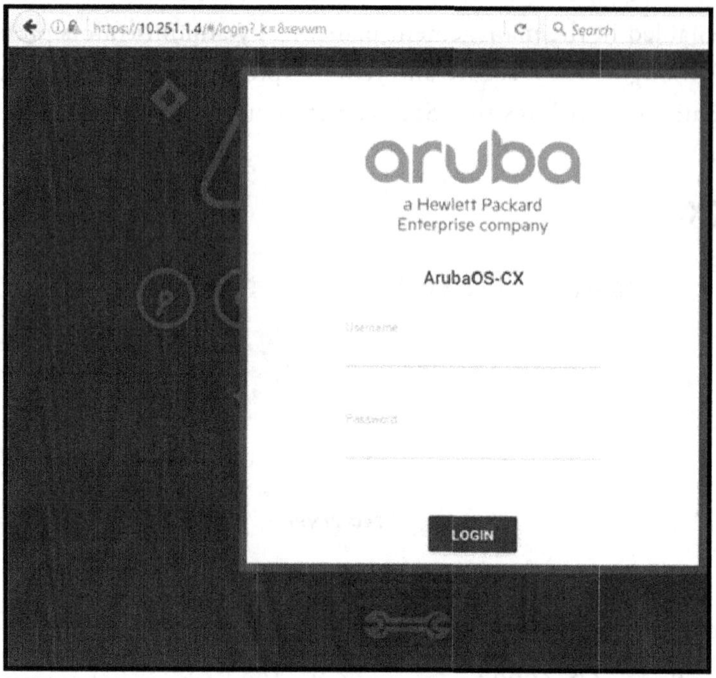

Figure 12-5: Login page of the web interface

The landing page is as following:

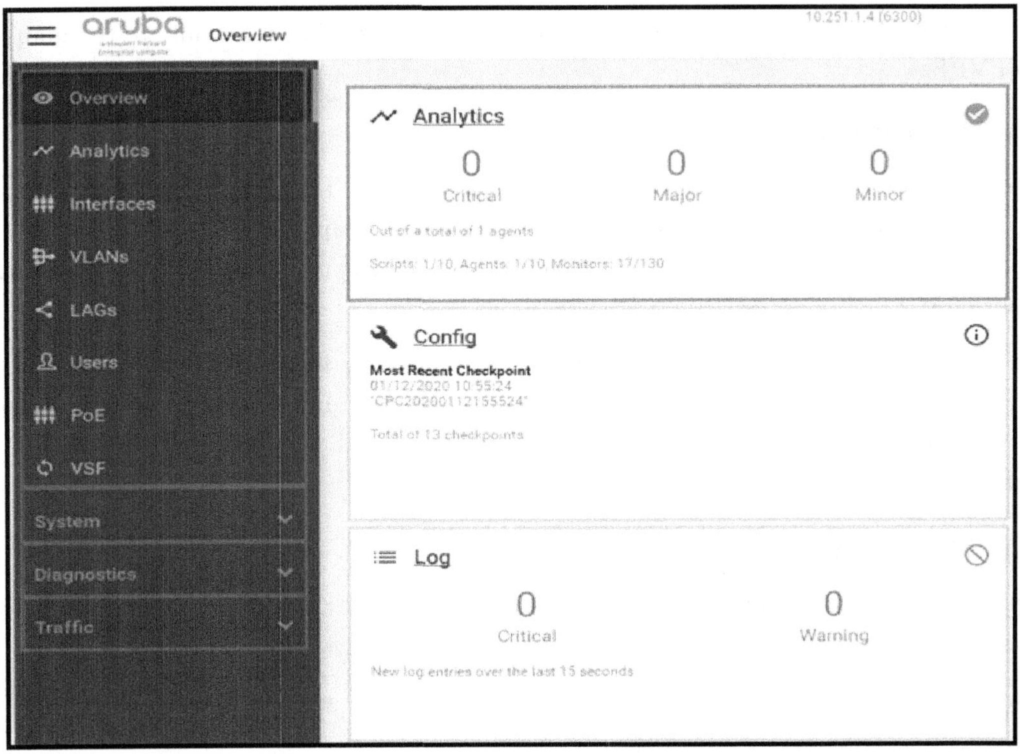

Figure 12-6: Landing page of the web interface

The menu on the left, highlighted in red, is divided into four different sections:

Overview

Displays basic protocols and feature information. Those include analytics, Interfaces, VLANs, LAGs, users, Power over Ethernet (PoE), and VSF. The analytics section gives you access Network Analytics Engine (NAE).

System

Displays environmental information, log, DNS servers, SNMP configuration, and configuration management.

Diagnostics

Includes features like ping and traceroute.

Traffic

Includes information about spanning tree.

Remember that it is a best practice to permit HTTPS/GUI by using them only in the management VRF.

Administrative RBAC

Administrators can access and configure the switch as required. They have full visibility of all switch processes. This is perfect for well-trained, trustworthy employees, but others should be restricted based on their expertise and job function. This concept is known as Role Based Access Control (RBAC).

AOS-CX applies RBAC defining user groups, where each group is manually configured with specifically allowed switch commands. AOS-CX defines three default user groups: administrators, auditors, and operators, which can neither be deleted nor edited. You can use the `show user-group` command to verify this information.

Built-in user groups and their privileges

The switch provides the following built-in user groups with corresponding roles. Each of these roles comes with a set of privileges.

Administrators

Administrators have full privileges, including:

- Full CLI access.
- Performing firmware upgrades.
- Viewing switch configuration information, including sensitive information such as passwords which are displayed as ciphertext.
- Performing switch configuration.
- Adding/removing user accounts.
- Configuring users accounts, including passwords. Once set, a password cannot be deleted or set to empty.
- REST API: All methods (GET, PUT, POST, DELETE) and switch resources are available.

The privilege level for administrators is 15.

Auditors

Auditors are restricted to functions related to auditing only:

- CLI: Access to commands in the auditor context (auditor>) only.
- Web UI: Access to the System > Log page only.

- REST API: POST method available for the \login and \logout resources. GET method available for the following resources only:
    - Audit log: /logs/audit
    - Event log: /logs/event

The privilege level for auditors is 19.

Operators

Operators have no switch configuration privileges. Operators are restricted to:

- Basic display-only CLI access.
- Viewing of nonsensitive switch configuration information.
- REST API: Other than the \login and \logout resources, only the GET method is available.

The privilege level for operators is 1.

Configuration

The following are commands to configure RBAC.

- Define new user groups:

```
Switch(config)# user-group <GROUP-NAME>
```

- Map a user account to a defined group:

```
Switch(config)# user <USERNAME> group <USER-GROUP> password plaintext
<PASSWORD>
```

- Use rules to define the commands that will be available:

```
Switch(config)# user-group monitoring
Switch(config-usr-grp-monitoring)# permit cli command "<COMMAND>"
```

In the following example, a user group named "monitoring" can only use the `show version` and `show interface 1/1/1` commands.

```
Switch(config)# user-group monitoring
Switch(config-ur-grp-monitoring)# 10 permit cli command "show version"
Switch(config-ur-grp-monitoring)# 20 permit cli command "show interface 1/1/1"
```

The user group rule evaluation proceeds from the lowest to the highest sequence number until the first successful match, resulting in either CLI command permission or denial. Rule evaluation ceases upon the first match. Therefore, rules for related CLI commands must be defined in most restrictive to least restrictive order. When creating the rules, a best practice is to leave a space in between rules to allow the network administrator to create new rules in a higher position and give some space to move rules as needed.

RADIUS-based management authentication

For authentication, some administrators define local credentials on each switch. This is good for smaller environments but becomes a burden as the network grows. If you have 100 network devices, you must maintain 100 separate credential databases, and try to keep them consistent. Many people prefer to centralize AAA services on a Remote Authentication Dial-In User Service (RADIUS) server. Hundreds and thousands of devices can use the RADIUS protocol (UDP port 1812) to access a centralized RADIUS server – a centralized repository for credentials.

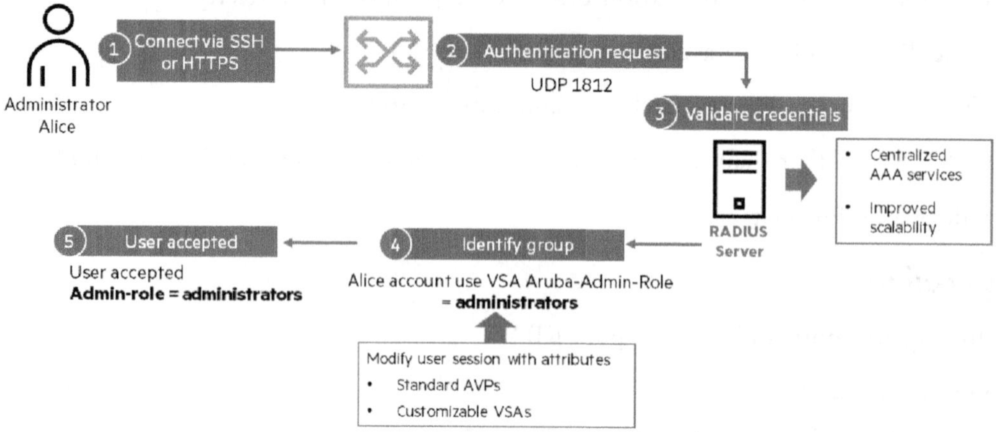

Figure 12-7: Steps involved in RADIUS-based management authentication

A user or administrator asks the switch if it can connect, and the switch in turn asks the RADIUS server: "Hey, I have user Alice, with password Secret123. Should I grant them access?"

The RADIUS server receives this request, validates credentials against its database, and responds to the switch with an ACCEPT message or a REJECT message. RADIUS messages can also include attributes. These are like "notes" that can modify how a connection session shall proceed.

For example, user Alice attempts to login to a switch, which sends Alice's credentials to the RADIUS server. The server validates these credentials and sends an ACCEPT message, which can include an attribute that helps the AOS-CX switch to place Alice in the Administrator user group.

One reason for the RADIUS standard's popularity is that it was designed to be customizable by equipment manufacturers and vendors, using Vendor Specific Attributes (VSAs). HPE has created VSAs that are specific for its products, for example, Aruba VSAs and Hewlett Packard Enterprise VSAs, to enhance usability and features. By default, RADIUS servers use standard attributes called Attribute Value Pairs (AVPs). Custom VSAs must be manually added to the RADIUS server. ClearPass is a powerful RADIUS server that has VSA dictionaries for many vendors, including HPE and Aruba, without need for manual addition.

Simple Network Management Protocol

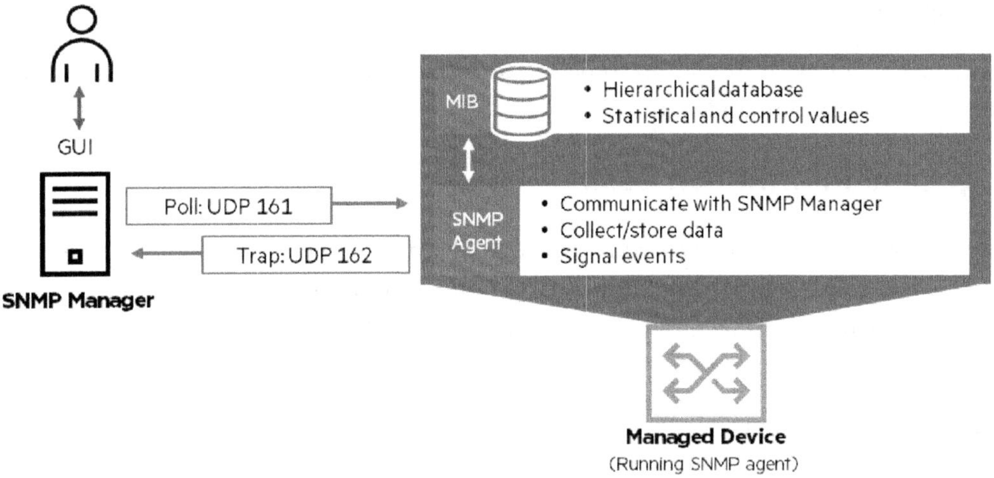

Figure 12-8: SNMP architecture

Simple Network Managment Protocol (SNMP) is an application-layer protocol defined in RFC 1157 for exchanging management information between network devices. This protocol is a popular method to configure, manage, and monitor network elements from any vendor. The figure shows the SNMP architecture, as follows:

SNMP manager

Is typically a dedicated server that communicates with the SNMP agent. It often has a GUI interface, so administrators can view and change parameters for many network devices from a single interface. The manager uses several message types to gather and modify network device parameters. Some of its key functions include:

- Queries agents
- Gets responses from agents
- Sets variables in agents
- Acknowledges asynchronous events from agents

Managed devices

These include the routers, switches, firewalls, wireless controllers, access points, and any other network devices that should be centrally monitored, managed, and configured. Internally, these managed devices have two main SNMP components.

SNMP agent

The SNMP agent allows information to be collected and communicated to the SNMP manager about device configuration and status.

Management Information Base

Management Information Base (MIB) is a database used in networking to manage and monitor network devices. It is a standardized framework defined by the SNMP that allows network administrators to gather information about network devices, monitor their performance, and control their behavior.

Communication and ports

Usually, SNMP manager to agent communication is a "polling" mechanism. When this device requires to get information from the managed device, it requests the information using simple commands such as GET, GET NEXT, and GET BULK. Changes in the device are done using the SET command. This communication uses UDP 161.

The exception to the polling mechanism is when the managed device has important information that cannot wait for the manager's period polling. For example, "Newsflash! My CPU utilization is dangerously high." In cases like this, the managed device "pushes" information to the manager without being asked, using an SNMP Trap. SNMP Traps use UDP 162.

SNMP versions

Let us now look at the various SNMP versions available.

- **SNMP version 1:** The first version of the SNMP protocol uses a community-based security mechanism. The "community string" is like a simple preshared passcode. If the managed device has a community string of public, then any SNMP manager that can reach the device can access the agent's MIB, as long as `community string = public` is used.
- **SNMP version 2c:** This is the revised protocol, which includes enhancements of SNMPv1 in the areas of protocol packet types, transport mapping and MIB structure elements. This version also uses a community-based security mechanism.

  A simple configuration using SNMP version 2 in AOS-CX is presented in the following lines:

  ```
  Switch(config)# snmp-server community <COMMUNITY-STRING>
  Switch(config)# snmp-server vrf <VRF-NAME>
  ```

- **SNMP version 3:** Provides a higher level of security, including message integrity, authentication, and encryption.

By default, SNMP read-only community is set to public in AOS-CX. The write-only community is not set, therefore it is not accessible. AOS-CX switches are read-only by default.

Configuration file management

An AOS-CX switch uses RAM and flash memory to save configuration files. It is important to understand RAM and flash memory characteristics, as they relate to the storage of device configuration files.

Figure 12-9: Topology highlighting key elements of RAM memory and flash memory

RAM memory is used to save the configuration that is currently running on the device, the running-configuration. Remember that RAM is not permanent storage, the RAM contents are lost when a device reboots or powers down. If you modify the running-configuration and the switch reboots, your changes are lost. Use `show running-config` command to view this file.

Flash memory is used to permanently save a configuration file. The contents of flash memory survive during reboot or power off. Use the `show startup-config` command to verify the contents on this file.

Of course, any time you make changes to the running-configuration, you should copy those changes to the startup-config using the copy command. To permanently save the configuration that is running in the device, use the `copy running-config startup-config` command.

Alternatively, you can use the `write memory` command, which does the exact same task.

If you need to restore a configuration from the startup-configuration use the `copy startup-config running-config` command.

As a best practice, you should export the running and the startup configuration files to an external file server. This will help you to recover from catastrophic device failures. To accomplish this task, you can enter the `copy` command, as follows:

```
Switch# copy {startup-config | running-config}  {tftp://|sftp://USER@}{IP|HOST}
[:PORT][;blocksize=VAL]/FILE
```

For example, to copy the running-configuration to a TFTP server with IP address 10.253.1.21, use the `copy running-config tftp://10.253.1.21/switch_config.cfg` command.

AOS-CX switches also allow you to copy files to a USB flash drive, which can be directly connected to the switch. To copy your running or startup-configuration to a USB device, enter the following syntax:

```
Switch# copy {running-config | startup-config} usb:/file
```

Maintenance

Maintenance helps you to ensure all key secure management elements are working as desired, and it also helps you to troubleshoot if required. We are going to look at some of the key elements involved in maintenance.

Overview of checkpoint

Using the `copy` commands, you learned that the switch allows you to store configuration files and recover lost configurations in case of a power outage or administrative mistake. This is a good thing, but these files do not include any additional data about the state of networking processes. For a true recovery, a new approach is needed. A checkpoint is a snapshot of the switch running configuration and its relevant metadata during the time of creation. You can use checkpoints to apply the switch configuration stored within a checkpoint whenever needed, such as reverting to a previous working configuration. A checkpoint is flexible since it can be applied to other switches of the same model.

There are two types of checkpoints:

- **System-generated** checkpoints are generated after five minutes of inactivity after a configuration change. By default, a checkpoint is taken every 300 seconds, or five minutes. The format that this checkpoint type uses is CP<YYYYMMDDHHMMSS>, for example: CPC20200111000038.
- **User-generated** checkpoints are manually generated by the administrator.

Configuring checkpoint

To create a checkpoint, enter the following `copy` command:

```
Switch# copy {running-config | startup-config} checkpoint <CHECKPOINT-NAME>
```

To verify the checkpoint list, enter the `show checkpoint` command.

To revert the switch from a checkpoint enter the following command:

```
Switch# copy checkpoint <CHECKPOINT-NAME> {running-config | startup-config}
```

To erase a checkpoint, enter the following command:

```
switch# erase checkpoint <CHECKPOINT-NAME>
```

To erase the startup configuration, enter the following command:

```
switch# erase startup-config
```

To erase all checkpoints, enter the following command:

```
switch# erase checkpoint all
```

> You can export a checkpoint file to an external server.

Alternatively, to rollback to a checkpoint, you can use the `checkpoint rollback` *<CHECKPOINT-NAME>* command. This command only restores the configuration to the running-config file.

> The `checkpoint rollback` command does the same task as the `copy checkpoint running-config` command.

Checkpoint auto mode

There are situations when applying a configuration could lock out your access to the switch. This leaves no other option but to locally connect to the device using the console port and remove the offending configuration lines.

The checkpoint auto mode feature addresses this problem in an elegant way. The solution includes a timer that you set. The system then creates a checkpoint with the current state; you can then make your desired configuration changes, which the system applies, after a confirmation. If the configuration lines did not affect your access, you will be able to confirm the change. If your changes locked you out of the device, confirmation does not occur. The switch thus automatically reverts to the latest snapshot after the time expires.

To use this mode, enter the following command:

```
Switch# checkpoint auto <MINUTES>
Switch# checkpoint auto confirm
```

Operating system image management introduction

AOS-CX switches have two flash memory modules or partitions for storing switch software image files. AOS-CX refers to those modules as primary and secondary images.

Having two flash memory locations helps to manage the switch more easily and plan firmware upgrades. For example, you can download a new firmware code in the flash memory that is not currently in use and then instruct the switch to boot from it during a maintenance window.

The `show images` command shows information about the software in the primary and secondary images. A sample output is as follows:

```
Switch# show images
---------------------------------------------------------------
AOS-CX Primary Image
---------------------------------------------------------------
Version : GL.10.04.0003
```

```
Size      : 415 MB
Date      : 2019-10-31 12:33:52 PDT
SHA-256 : 794336ee0b1f941aac421bec5333f16b4ebbcd1f4035a388d3b79aa
------------------------------------------------------------------
AOS-CX Secondary Image
------------------------------------------------------------------
Default Image : primary
------------------------------------------------------------
Management Module 1/1 (Active)
------------------------------------------------------------
Active Image         : primary
Service OS Version : GL.01.05.0002
BIOS Version         : GL-01-0013
```

Operating system image management access

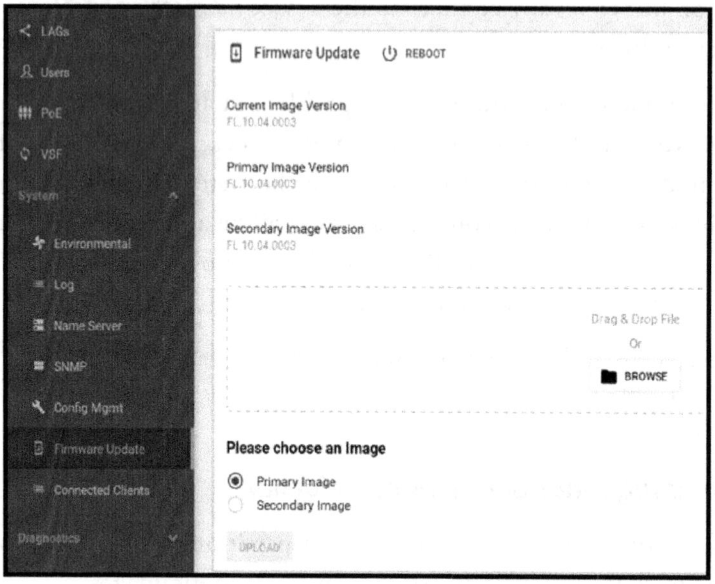

Figure 12-10: The Firmware Update tab

You can update the switch using the CLI or the GUI. Using the GUI, you have to select the **Firmware Update** tab as shown in the figure. You can either use the **Drag & Drop File** option or browse for the image and upload it.

Using the CLI, enter the `copy` command as follows:

```
Switch# copy {tftp://|sftp://USER@}{IP|HOST}[:PORT][;blocksize=VAL]/FILE {primary
 | secondary}
```

Suppose that you want to copy the image file GL.10.04.0003.swi from the TFTP server with IP address 10.253.1.21 to the Primary partition. Use the following command:

```
Switch# copy tftp://10.253.1.23/GL.10.04.0003.swi primary
```

To use Secure FTP with the username "admin", enter the following command:

```
Switch# copy sftp://admin@10.253.1.21/GL.10.04.0003.swi primary
```

> HPE Aruba Networking Central can be used to update the operating system.

Password recovery process

The steps to recover the password, if required, are as follows:

1. Connect to the switch using the console port.
2. Power cycle the switch.
3. When the system prompts, select the **Service OS** console option by typing **0**.

```
Boot Profiles:
0. Service OS Console
1. Primary Software Image [FL.10.04.0003]
2. Secondary Software Image [FL.10.04.0003]
Select profile(primary): 0
```

4. Log in with the username "admin". No password is set for this account.
5. Enter the password keyword and type the new password for the admin account.

```
ServiceOS login: admin
SVOS> password
Enter password: ********
Confirm password: ********
```

6. Enter `boot`.
7. Log in using the admin username and the password that was set in step 5.

Learning check

Check your understanding of this material by answering the following questions.

Question 1

What is true about managing AOS-CX devices? (Select two)

- ☐ AOS-CX switches do not support out-of-band management (OOBM).
- ☐ AOS-CX switches starting with the CX 6200 Series and up support OOBM.
- ☐ In-band management and out-of-band management are mutually exclusive.
- ☐ In-band management is often used in campus networks.

Question 2

What are the default AOS-CX user groups? (Select three)

- ☐ Manager
- ☐ Administrator
- ☐ Visitor
- ☐ Operator
- ☐ Guest
- ☐ Auditor

13 Resources

The following resources can help in your understanding of implementing switching with the help of the HPE Aruba Networking CX Switch Operating System.

HPE Networking Support Portal

This is a 'one stop shop' for all things HPE Aruba Networking, including software, documentation, license management, and technical support.

It provides the following key features among others:

- Management of customer service requests
- Wide range of software and document resources
- Notifications for updates on products and services
- Tracking subscription details
- Links to additional support and resources

Access the ASP at https://networkingsupport.hpe.com.

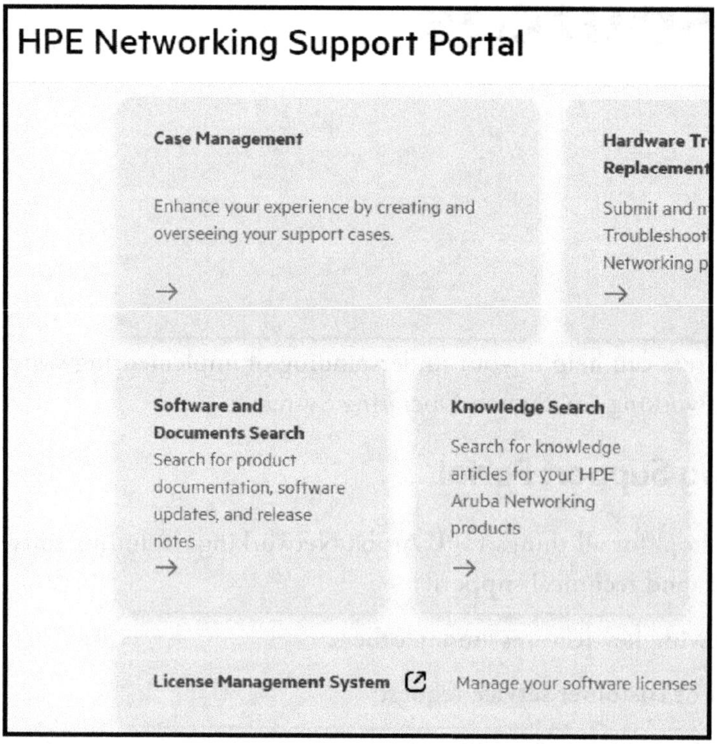

Figure 13-1: HPE Networking Support Portal dashboard

Validated Solution Guides (VSG)

VSGs provide you access to network designs to rapidly deploy HPE Aruba Networking solutions. They offer cross-portfolio solution guides describing the implementations of primary customer use cases. They are validated by HPE's Solution TME and Solution Quality Assurance teams on an ongoing basis using a rigorous process. They not only provide prescriptive guidance focused on the HPE recommended best practices specific to the solution being covered, but also enable customers and partners to install end-to-end solutions.

Access the VSGs at https://www.arubanetworks.com/techdocs/VSG/.

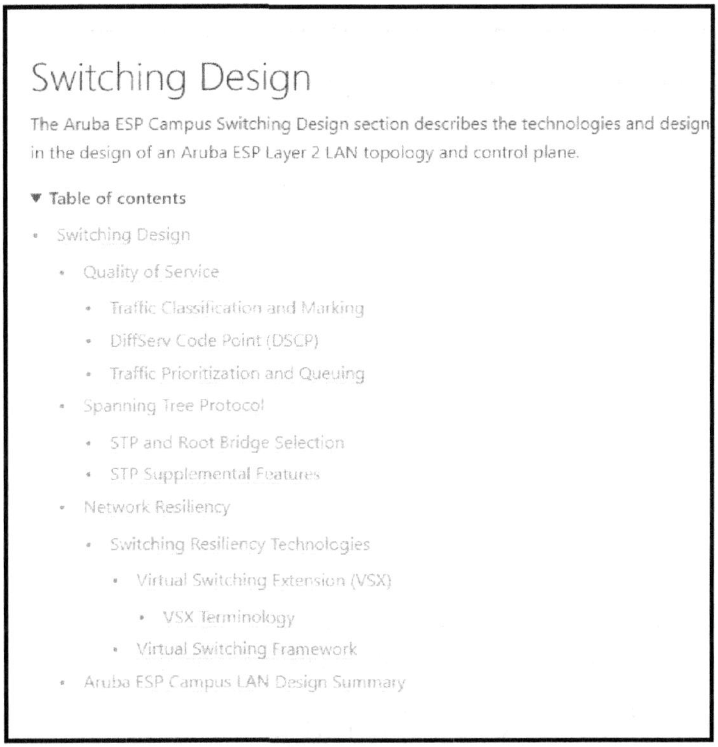

Figure 13-2: Validated solution guides example

Airheads community

At the Airheads community, you will find the latest news and announcements, discussion boards, community learning, events, and support. Get registered and become an Airhead today! Access the community page at https://www.arubanetworks.com/techdocs/VSG/.

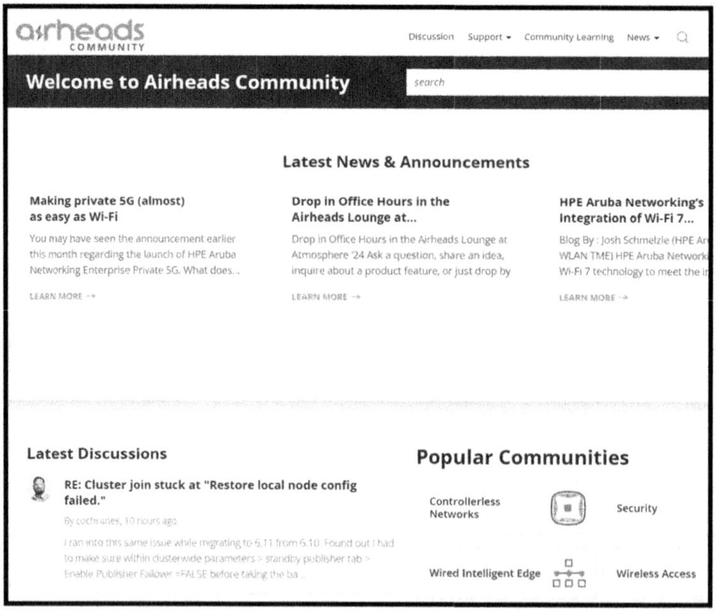

Figure 13-3: Airheads community dashboard

HPE Aruba Networking Solution Exchange

A community managed solution repository which includes:

- Step-by-step configuration support through an interactive wizard
- Solutions that are supported on multiple platforms
- Code validation features

Access the ASE at https://www.arubanetworks.com/techdocs/VSG/.

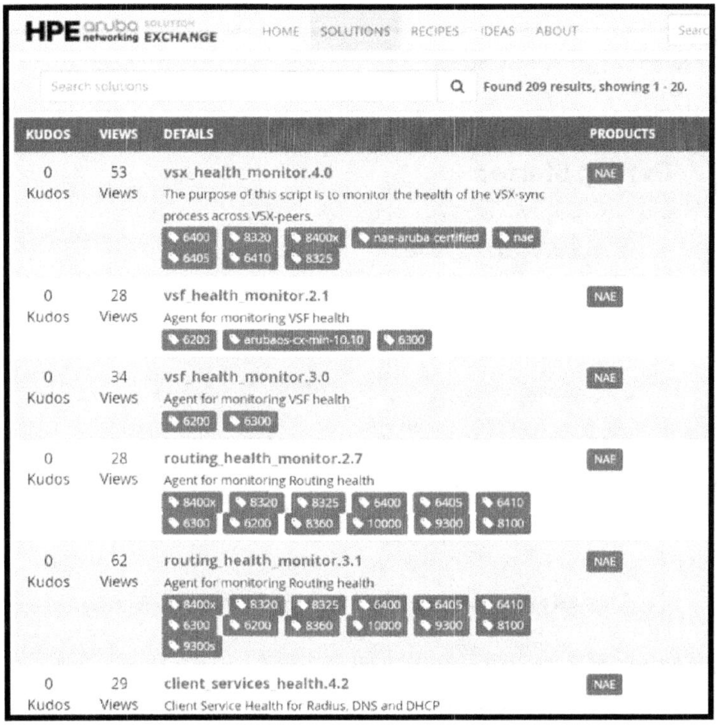

Figure 13-4: Sample list of solutions from HPE Aruba Networking Solution Exchange

HPE Aruba Networking Developer Hub

HPE Aruba Networking Developer Hub is where to find help and resources to automate networks and quickly build powerful applications by integrating with HPE Aruba Networking products. The hub offers a curated repository of learning content including code samples, APIs, and user guides.

Access the Developer Hub at https://www.arubanetworks.com/techdocs/VSG/.

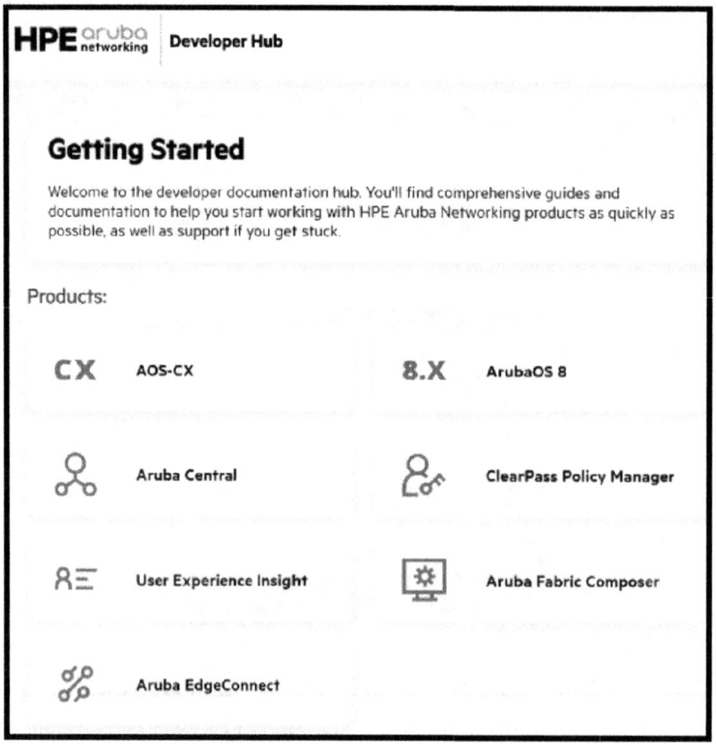

Figure 13-5: Developer Hub "Products" dashboard

14 HPE6-A86 practice exam

This chapter provides sample test questions based on the style and difficulty you would find on the real HPE6-A86 exam for the HPE Aruba Networking Switching Associate certification.

OBJECTIVES

The purpose of this practice exam is to:

- ✓ Reinforce the material you have learned.
- ✓ Become familiar with the type and style of questions found on the HPE Aruba Networking Switching Associate exam.

Exam overview

This exam tests a candidate's understanding of wired technologies used in SMB, edge, and simple core environments.

Ideal candidate for this exam

Typical candidates are HPE customers, partners, and employees with at least six months of experience working as a junior team member who supports and maintains wired networks or works independently on limited-scope projects.

Examples of job roles include: Network operator, tier 1/2 support, support technician, field technician, and network administrator.

Exam contents

This exam has 75 questions. Here are the types of questions to expect:

- Multiple choice (multiple responses)
- Multiple choice (single response)
- Drag-and-drop

Advice to help you take this exam

- Complete the training and review all materials and documents before you take the exam.
- Exam items are based on expected knowledge acquired from job experience, an expected level of industry-standard knowledge, or other prerequisites (events, supplemental materials, and so forth).
- Successful completion of the course or study materials alone does not ensure you will pass the exam.

Read the entire question and consider all options before you answer. If the question includes an exhibit, study the exhibit and read the question again. Select the answer that fully responds to the question. If the question asks for more than one answer, select all correct answers. There is no partial credit.

Exam objectives and exam information

This section covers the objectives for the HPE6-A86 exam, as well as information about the exam.

HPE6-A86 exam objectives

Table 14-1 breaks down the objectives covered on the exam.

Table 14-1: HPE6-A86 exam objectives

Percentage of exam	Sections/Objectives
24%	1. Identify, describe, and apply foundational networking architectures and technologies. a. Describe and explain the OSI Model. (2%) b. Describe and explain the most common layer media (Layer 1). (2%) c. Describe the basics of Layer 2 Ethernet, including broadcast domains. (3%) d. Describe the basics of Layer 3 addressing, including subnetting, ARP, and routing. (4%) e. Define and recognize the purpose and interaction of Layer 4 (Transport) protocols in an IP network. (1%) f. Identify the role of various network management protocols such as TFTP, SFTP, FTP, Telnet, and SNMPv2 (2%) g. Identify and describe the concept of QoS and explain its significance in converged networks. (1%) h. Describe and explain the basic network security setup on HPE Aruba Networking switches. (2%) i. Describe Layer 2 redundancy technologies such as STP/RSTP and VSF, including their benefits. (2%) j. Describe and apply link aggregation. (2%) k. Identify, describe, and explain VLANs (3%)
21%	2. Identify, describe, and differentiate HPE Aruba Networking products' and solutions' functions and features. a. Identify basic features and management options for HPE Aruba Networking wired products (5%) b. Identify the software capabilities and differentiators in the HPE Aruba Networking switching product line (6%) c. Identify the hardware capabilities and differentiators in the HPE Aruba Networking switching product line (5%) d. Identify and describe available tool sets for managing HPE Aruba Networking products (CLI-based, web, scripted, SNMP, HPE Aruba Networking Central, mobile app, and API). (5%)

Percentage of exam	Sections/Objectives
22%	3. Install, configure, set up, and validate wired network solutions. a. Configure basic features on HPE Aruba Networking switches, including initial settings and management access. (7%) b. Configure HPE Aruba Networking switches with Layer 2 technologies such as RSTP/MSTP, link aggregation, VLANs, and LLDP. (6%) c. Configure HPE Aruba Networking switches with Layer 3 technologies including IP addressing, routing (5%) d. Validate the installed solution via show commands. (4%)
17%	4. Troubleshoot, operate, and maintain HPE Aruba Networking solutions. a. Troubleshoot switched and routed networks. (10%) b. Use general troubleshooting tools. (7%)
16%	5. Manage and monitor HPE Aruba Networking solutions. a. Perform network management and administrative tasks according to best practices. (16%)

Exam information

Table 14-2 summarizes the information to register for the exam. You need an HPE Learner ID and a Pearson Vue login and password. No reference material is allowed at the testing site. This exam may also contain beta test items for experimental purposes. During the exam, you can make comments about the exam items. HPE welcomes these comments as part of our continuous improvement process. To register for the exam, visit https://www.pearsonvue.com/hpe.

Table 14-2: HPE6-A86 exam information

Exam ID	HPE6-A86
Exam type	Proctored
Exam duration	2 hours
Exam length	75 questions
Passing score	68%
Delivery languages	English
Supporting resources	HPE Aruba Networking Certified Associate - Switching, Rev. 24.31 course, available in physical, virtual, and eLearning formats.

More and current information concerning the HPE6-A86 exam can be found here: https://certification-learning.hpe.com/tr/datacard/exam/HPE6-A86.

Practice exam questions and answers

The following questions will help you to measure your understanding of topics that may be covered in the exam. Read all the choices carefully since there may be more than one correct answer. Choose all correct answers for each question.

Questions

1. In which layer should the network have the faster switches (throughput)?

    A. Access layer

    B. Aggregation layer

    C. Core layer

    D. Layer 2

2. Which switches provide always on PoE and Quick PoE? (Select three)

    A. CX 6300 Series

    B. CX 6400 Series

    C. CX 8360 Series

    D. CX 4100i Series

3. Which AOS-CX switch series are chassis based? (Select two)

    A. CX 8325 Series

    B. CX 6400 Series

    C. CX 10000 Series

    D. CX 8400 Series

4. You have a small enterprise network with a limited budget and a small IT department. Which network design type is best?

    A. A three-tier design.

    B. A two-tier design.

    C. A campus network design.

    D. A small data center network.

5. Which are valid methods for provisioning AOS-CX switches? (Select three)
    A. Manual provisioning
    B. Zero-Touch Provisioning
    C. AirWave
    D. HPE Aruba Networking IMC
    E. Fabric Composer
6. In order for the ZTP process to succeed, which of the following are required?
    A. HPE Aruba Networking Central
    B. A DHCP server with options 43, 66, and 67
    C. A DHCP server with options 5, 23, and 67
    D. The HTTP Proxy Location
7. Which options accurately describe MAC address and ARP tables? (Select two)
    A. Switches automatically build the MAC table based on destination MAC addresses.
    B. MAC address tables are built based on source IP addresses.
    C. The ARP reply packet uses a broadcasting mechanism.
    D. The ARP table maps IP addresses to MAC addresses.
    E. Switches use the MAC address table to properly forward frames.
8. When might a switch add a VLAN tag to a frame?
    A. When it forwards the frame to a server
    B. When it forwards the frame to a PC or client
    C. When it forwards the frame to another switch
    D. When it accepts a frame from another switch
    E. When it accepts a frame from a server
9. In RSTP, how is the root bridge chosen in a network topology?
    A. Based on the lowest MAC address of a switch.
    B. Based on the highest MAC address of a switch.
    C. Based on the lowest BID of a switch.
    D. Based on the highest BID of a switch.

10. Which port role in RSTP is responsible for forwarding traffic toward the root bridge.
    A. Root port
    B. Designated port
    C. Alternate port
    D. Edge port
11. How does dynamic Link Aggregation Group (LAG) differ from a static LAG?
    A. Dynamic LAGs require manual configuration by network administrators.
    B. Static LAGs utilize protocols like LACP for automatic link aggregation.
    C. Static LAGs provide load-balancing capabilities without additional configuration.
    D. Dynamic LAGs automatically negotiate and manage link aggregation.
12. Which load balancing algorithm typically uses source and destination IP addresses to determine traffic distribution in a LAG?
    A. Round Robin
    B. Hash-based
    C. Weighted
    D. Least Connections
13. What is the primary purpose of implementing a LAG in a network environment?
    A. To increase the number of network devices connected to the network.
    B. To reduce the overall cost of network infrastructure deployment.
    C. To prioritize specific types of traffic over others.
    D. To aggregate multiple physical links into a single link for increased bandwidth and redundancy.
14. What is the primary purpose of VSF configuration?
    A. To distribute network traffic across multiple physical switches.
    B. To manage network devices remotely.
    C. To combine multiple physical switches into single virtual chassis.
    D. To optimize network routing protocols.
15. How does VSF enhance network resilience?

A. By combining multiple physical switches into a single logical entity.

B. By providing active-active redundancy and seamless failover.

C. By encrypting network traffic for enhanced security.

D. By increasing network bandwidth through link aggregation.

16. Given IP address 172.20.3.54 and a mask of 255.255.255.0, what can be accurately stated about this addressing? (Select three)

    A. The host portion of the address is .3.54.

    B. The network portion of the address is 172.20.3.

    C. Host 172.20.3.89 would be on the same network.

    D. A switch might use this address to forward the packet.

    E. You could also indicate this mask as "/24".

17. Which aspect of Layer 3 analysis involves evaluating the operation and effectiveness of protocols such as RIP, OSPF, EIGRP, and BGP in dynamically exchanging routing information and building routing tables?

    A. Routing table analysis

    B. Routing convergence

    C. Route redistribution

    D. Routing protocols

18. Which type of interface is used for internal routing and connects a Layer 2 VLAN to the internal routing functions in a multi-layer switch.

    A. Trunk port

    B. Layer 2 physical routed port

    C. Switch Virtual Interface (SVI)

    D. Layer 2 switch port

19. What is the purpose of a routing table in IP networking, and how does it influence the forwarding of data packets?

    A. It stores information about the next-hop routers.

    B. It determines the best path for forwarding data packets.

   C. It defines the subnet masks for connected networks.

   D. It regulates the flow of traffic within a network segment.

20. Which type of routing protocol dynamically adapts to changes in network topology and link status?

   A. Static routing

   B. Dynamic routing

   C. Distance-vector routing

   D. Link-state routing

21. What is the default multicast IP address used for sending OSPF hello packets?

   A. 224.0.0.1

   B. 224.0.0.2

   C. 224.0.0.3

   D. 224.0.0.4

   E. 224.0.0.5

22. Which phase of OSPF involves running the SPF algorithm to determine the best path for each network destination?

   A. Phase 1: Build a Neighbor table

   B. Phase 2: Build a topology database

   C. Phase 3: Build OSPF route table

   D. Phase 4: Build the route table

23. How does AOS-CX determine the Router Identifier (RID) if it is not manually specified?

   A. It uses the IP address of the loopback interface with the lowest IP address.

   B. It uses the IP address of the loopback interface with the highest IP address.

   C. It uses the IP address of the non-loopback interface with the lowest IP address.

   D. It uses the IP address of the non-loopback interface with the highest IP address.

   E. It randomly generates a unique RID.

24. What is the definition of jitter in networking and communication systems?

   A. The delay variation in packet arrival times.

   B. The measure of speed of data transmission.

C. The maximum bandwidth of a network connection.

D. The number of errors in data transmission.

25. Which of the following best describes the purpose of rate limiting in networking?

    A. Maximize network bandwidth

    B. Preventing unauthorized access

    C. Minimizing latency

    D. Ensuring fair distribution of resources

26. What does DSCP stand for and at which layer of the OSI model does it operate?

    A. Data Service Control Point, Layer 2

    B. Differentiated Service Code Point, Layer 2

    C. Differentiated Service Code Point, Layer 3

    D. Data Segment Control Protocol, Layer 3

27. Which access control method involves granting access to network resources based on the physical address of the Network Interface Card (NIC)?

    A. Port authentication

    B. Secure certificate verification

    C. Web authentication

    D. MAC authentication

28. Which of the statements describe valid AOS-CX secure management access concepts? (Select two)

    A. AAA accounting controls what you can do once you log in to the device.

    B. AAA services can be used in both the management VRF and the default VRF.

    C. Role-Based Access Control (RBAC) is based on a user's group assignment.

    D. Advantages of using a RADIUS server include simplicity and faster authentication operations.

    E. An easy way to modify several users' access is to modify the "operators" group.

29. What is true about AOS-CX configuration file management? (Select three)

A. It is a good idea to copy the running configuration to the startup configuration after you have made changes to the configuration file that is currently in use.

B. The configuration file that is used while the switch is in operation is typically stored in a flash file system.

C. The advantage of a "checkpoint" is that it stores operational metadata, in addition to the device configuration.

D. Checkpoints are files—a snapshot of the configuration at a given time.

Answers

1. In which layer should the network have the faster switches (throughput)?

   A. Access layer

   B. Aggregation layer

   C. Core layer

   D. Layer 2

   Answer **C** is correct. At the core layer, you want the throughput to be fast to move packets.

2. Which switches provide always on PoE and Quick PoE? (Select three)

   A. CX 6300 Series

   B. CX 6400 Series

   C. CX 8360 Series

   D. CX 4100i Series

   Answers **A**, **B**, and **D** are correct. Only certain access switches provide PoE. The CX 8360 Series is classified as a core or data center switch.

3. Which AOS-CX switch series are chassis based? (Select two)

   A. CX 8325 Series

   B. CX 6400 Series

   C. CX 10000 Series

   D. CX 8400 Series

   Answers **B** and **D** are correct. The CX 6400 Series and CX 8400 Series are chassis switches.

4. You have a small enterprise network with a limited budget and a small IT department. Which network design type is best?

A. A three-tier design.

B. A two-tier design.

C. A campus network design.

D. A small data center network.

Answer **B** is correct. A two-tier design is most appropriate for a small enterprise network with a limited budget.

5. Which are valid methods for provisioning AOS-CX switches? (Select three)

A. Manual provisioning

B. Zero-Touch Provisioning

C. AirWave

D. HPE Aruba Networking IMC

E. Fabric Composer

Answers **A**, **B**, and **E** are correct. Manual provisioning, zero-touch provisioning, and Fabric Composer all are valid methods to provision switches.

6. In order for the ZTP process to succeed, which of the following are required?

A. HPE Aruba Networking Central

B. A DHCP server with options 43, 66, and 67

C. A DHCP server with options 5, 23, and 67

D. The HTTP Proxy Location

Answer B is correct. The ZTP process requires a DHCP server configured with options 43, 66, and 67. Both HPE Aruba Networking Central and an HTTP Proxy Location are optional.

7. Which options accurately describe MAC address and ARP tables? (Select two)

A. Switches automatically build the MAC table based on destination MAC addresses.

B. MAC address tables are built based on source IP addresses.

C. The ARP reply packet uses a broadcasting mechanism.

D. The ARP table maps IP addresses to MAC addresses.

E. Switches use the MAC address table to properly forward frames.

Answers **D** and **E** are correct. The ARP table maps IP addresses to MAC addresses. Switches use the MAC address table to properly forward frames.

8. When might a switch add a VLAN tag to a frame?

   A. When it forwards the frame to a server

   B. When it forwards the frame to a PC or client

   C. When it forwards the frame to another switch

   D. When it accepts a frame from another switch

   E. When it accepts a frame from a server

   Answer **C** is correct. A switch might add a VLAN tag to a frame when forwarding it to another switch, provided the connection between the switches is a trunk connection, AND the VLAN is allowed, AND the VLAN is not the native VLAN.

9. In RSTP, how is the root bridge chosen in a network topology?

   A. Based on the lowest MAC address of a switch.

   B. Based on the highest MAC address of a switch.

   C. Based on the lowest BID of a switch.

   D. Based on the highest BID of a switch.

   Answer **C** is correct. The root bridge is based on the lowest BID of a switch.

10. Which port role in RSTP is responsible for forwarding traffic toward the root bridge.

    A. Root port

    B. Designated port

    C. Alternate port

    D. Edge port

    Answer **A** is correct. The root port is responsible for forwarding traffic toward the root bridge.

11. How does dynamic Link Aggregation Group (LAG) differ from a static LAG?

    A. Dynamic LAGs require manual configuration by network administrators.

    B. Static LAGs utilize protocols like LACP for automatic link aggregation.

    C. Static LAGs provide load-balancing capabilities without additional configuration.

    D. Dynamic LAGs automatically negotiate and manage link aggregation.

Answer **D** is correct. Dynamic LAGs will automatically negotiate and manage link aggregation.

12. Which load balancing algorithm typically uses source and destination IP addresses to determine traffic distribution in a LAG?

    A. Round Robin

    B. Hash-based

    C. Weighted

    D. Least Connections

Answers **B** is correct. The algorithm is hash-based.

13. What is the primary purpose of implementing a LAG in a network environment?

    A. To increase the number of network devices connected to the network.

    B. To reduce the overall cost of network infrastructure deployment.

    C. To prioritize specific types of traffic over others.

    D. To aggregate multiple physical links into a single link for increased bandwidth and redundancy.

Answer **D** is correct. The primary purpose is to to aggregate multiple physical links into a single link for increased bandwidth and redundancy.

14. What is the primary purpose of VSF configuration?

    A. To distribute network traffic across multiple physical switches.

    B. To manage network devices remotely.

    C. To combine multiple physical switches into single virtual chassis.

    D. To optimize network routing protocols.

Answer **C** is correct. The primary purpose of VSF configuration is to combine multiple physical switches into single virtual chassis.

15. How does VSF enhance network resilience?

    A. By combining multiple physical switches into a single logical entity.

    B. By providing active-active redundancy and seamless failover.

    C. By encrypting network traffic for enhanced security.

    D. By increasing network bandwidth through link aggregation.

Answer **B** is correct. VSF enhances network resilience by providing active-active redundancy and seamless failover.

16. Given IP address 172.20.3.54 and a mask of 255.255.255.0, what can be accurately stated about this addressing? (Select three)

    A. The host portion of the address is .3.54.

    B. The network portion of the address is 172.20.3.

    C. Host 172.20.3.89 would be on the same network.

    D. A switch might use this address to forward the packet.

    E. You could also indicate this mask as "/24".

    Answers **B**, **C**, and **E** are correct. The network portion of the address is 172.20.3. Host 172.20.3.89 would also be on the same network and you can also indicate this mask as "/24".

17. Which aspect of Layer 3 analysis involves evaluating the operation and effectiveness of protocols such as RIP, OSPF, EIGRP, and BGP in dynamically exchanging routing information and building routing tables?

    A. Routing table analysis

    B. Routing convergence

    C. Route redistribution

    D. Routing protocols

    Answer **D** is correct. Routing protocols is the aspect of Layer 3 analysis that involves evaluating the operation and effectiveness of RIP, OSPF, EIGRP and BGP in dynamically exchanging routing information and building routing tables.

18. Which type of interface is used for internal routing and connects a Layer 2 VLAN to the internal routing functions in a multi-layer switch.

    A. Trunk port

    B. Layer 2 physical routed port

    C. Switch Virtual Interface (SVI)

    D. Layer 2 switch port

    Answer **C** is correct. Switch Virtual Interface (SVI) is used for internal routing and connects a Layer 2 VLAN to the internal routing functions in a multi-layer switch.

19. What is the purpose of a routing table in IP networking, and how does it influence the forwarding of data packets?

    A. It stores information about the next-hop routers.

    B. It determines the best path for forwarding data packets.

    C. It defines the subnet masks for connected networks.

    D. It regulates the flow of traffic within a network segment.

    Answers **B** and **D** are correct. The purpose is to determine the best path for forwarding data packets, to regulate teh flow of traffic within a network segment.

20. Which type of routing protocol dynamically adapts to changes in network topology and link status?

    A. Static routing

    B. Dynamic routing

    C. Distance-vector routing

    D. Link-state routing

    Answer **B** is correct. Dynamic routing adapts to changes in a network topology and link status.

21. What is the default multicast IP address used for sending OSPF hello packets?

    A. 224.0.0.1

    B. 224.0.0.2

    C. 224.0.0.3

    D. 224.0.0.4

    E. 224.0.0.5

    Answer **E** iscorrect. 224.0.0.5 is the default multicast IP address used for sending OSPF hello packets.

22. Which phase of OSPF involves running the SPF algorithm to determine the best path for each network destination?

    A. Phase 1: Build a Neighbor table

    B. Phase 2: Build a topology database

C. Phase 3: Build OSPF route table

D. Phase 4: Build the route table

Answer **C** is correct. Phase 3: Build OSPF route table determines the best path for each network destination.

23. How does AOS-CX determine the Router Identifier (RID) if it is not manually specified?

 A. It uses the IP address of the loopback interface with the lowest IP address.

 B. It uses the IP address of the loopback interface with the highest IP address.

 C. It uses the IP address of the non-loopback interface with the lowest IP address.

 D. It uses the IP address of the non-loopback interface with the highest IP address.

 E. It randomly generates a unique RID.

Answer **B** is correct. AOS-CX uses the IP address of the loopback interface with the highest IP address to determine the RID if it is not manually specified.

24. What is the definition of jitter in networking and communication systems?

 A. The delay variation in packet arrival times.

 B. The measure of speed of data transmission.

 C. The maximum bandwidth of a network connection.

 D. The number of errors in data transmission.

Answer **A** is correct. Jitter is the delay variation in packet arrival times.

25. Which of the following best describes the purpose of rate limiting in networking?

 A. Maximize network bandwidth

 B. Preventing unauthorized access

 C. Minimizing latency

 D. Ensuring fair distribution of resources

Answer **D** is correct. The purpose of rate limiting is for ensuring fair distribution of resources.

26. What does DSCP stand for and at which layer of the OSI model does it operate?

 A. Data Service Control Point, Layer 2

 B. Differentiated Service Code Point, Layer 2

   C. Differentiated Service Code Point, Layer 3

   D. Data Segment Control Protocol, Layer 3

   Answer **D** is correct. DSCP stands for Differentiated Service Code Point and operates on Layer 3 of the OSI model.

27. Which access control method involves granting access to network resources based on the physical address of the Network Interface Card (NIC)?

   A. Port authentication

   B. Secure certificate verification

   C. Web authentication

   D. MAC authentication

   Answer **D** is correct. MAC authentication involves granting access to network resources based on the physical address of the NIC.

28. Which of the statements describe valid AOS-CX secure management access concepts? (Select two)

   A. AAA accounting controls what you can do once you log in to the device.

   B. AAA services can be used in both the management VRF and the default VRF.

   C. Role-Based Access Control (RBAC) is based on a user's group assignment.

   D. Advantages of using a RADIUS server include simplicity and faster authentication operations.

   E. An easy way to modify several users' access is to modify the "operators" group.

   Answers **B** and **C** are correct. AAA services can be used in both the management VRF and the default VRF. Role-Based Access Control (RBAC) is based on a user's group assignment.

29. What is true about AOS-CX configuration file management? (Select three)

   A. It is a good idea to copy the running configuration to the startup configuration after you have made changes to the configuration file that is currently in use.

   B. The configuration file that is used while the switch is in operation is typically stored in a flash file system.

   C. The advantage of a "checkpoint" is that it stores operational metadata, in addition to the device configuration.

D. Checkpoints are files—a snapshot of the configuration at a given time.

Answers **A**, **C**, and **D** are correct. It is a good idea to copy the running configuration to the startup configuration after you have made changes to the configuration file that is currently in use. The advantage of a "checkpoint" is that it stores operational metadata, in addition to the device configuration. Checkpoints are files—a snapshot of the configuration at a given time.

Appendix A Learning check

Overview

This appendix contains the answers to all the Learning Check sections found in the previous chapters.

Chapter 1: AOS-CX switching portfolio

This section contains the Learning Check answers for Chapter 1.

Learning Check answers

Question 1: Data center networks are typically optimized for high bandwidth and low latency, while campus networks are designed to support a large number of users and devices over a broader geographic area.

Question 2: CX 6200M Series, and CX 6300 Series both allow for VSF.

Chapter 2: Switch fundamentals

This section contains the Learning Check answers for Chapter 2.

Learning Check answers

Question 1: USB and serial cable

Question 2: `show interface brief`

Question 3: LLDP is a Layer 2 discovery protocol and it is best to use LLDP after your network documentation is complete.

Chapter 3: VLANs

This section contains the Learning Check answers for Chapter 3.

Learning Check answers

Question 1: A routing device defines the edge of a broadcast domain. A switch port defines the edge of a collision domain.

Question 2: Smaller broadcast domains can improve performance. Separate VLANs mitigate risk.

Question 3: Native VLAN is VLAN-1.

Chapter 4: Spanning tree protocol

This section contains the Learning Check answers for Chapter 4.

Learning Check answers

Question 1: Loop prevention.

Question 2: It has a faster convergence time than STP and supports only a single spanning tree instance.

Question 3: It supports load balancing across multiple instances.

Chapter 5: Link aggregation

This section contains the Learning Check answers for Chapter 5.

Learning Check answers

Question 1: IEEE 802.3ad

Question 2: The aggregation of links is manually configured by network administrators.

Question 3: To distribute network traffic evenly across aggregated links.

Chapter 6: Switch stacking and extension

This section contains the Learning Check answers for Chapter 6.

Learning Check answers

Question 1: The data plane moves data from ingress to egress port, and the control and data planes are tightly integrated.

Question 2: Different models of the CX 6200 Series can be mixed in a VSF stack, and the use of VSF is recommended in the access layer of a campus network.

Question 3: You can daisy-chain or use ring topology in up to 10 VSF members, and the configuration of VSF may cause members to reboot.

Chapter 7: Layer 3 routing

This section contains the Learning Check answers for Chapter 7.

Learning Check answers

Question 1: The next Layer 3 device that should receive the packet.

Question 2: Using the IP helper feature in a multilayer switch and AOS-CX switches do not support hosts with dynamic IP addresses are correct statements.

Chapter 8: IP routing fundamentals

This section contains the Learning Check answers for Chapter 8.

Learning Check answers

Question 1: The router chooses the static route.

Question 2: They can be used to route packets inside an Autonomous System and they are more scalable than protocols like RIP.

Question 3: 0

Chapter 9: Dynamic IP Routing - Open Shortest Path First

This section contains the Learning Check answers for Chapter 9.

Learning Check answers

Question 1: Distance vector routing and Link State routing are not primary characteristics of OSPF.

Question 2: 89

Question 3: Point-to-Point Network

Chapter 10: Quality of Service

This section contains the Learning Check answers for Chapter 10.

Learning Check answers

Question 1: Layer 2

Question 2: Fair allocation of bandwidth based on weights.

Question 3: The delay variation in packet arrival times.

Chapter 11: Wired network security fundamentals

This section contains the Learning Check answers for Chapter 11.

Learning Check answers

Question 1: Confidentiality, availability, and integrity are the three basic objectives of network security.

Question 2: Authorization is the component of the AAA framework that involves checking what a user is allowed to do.

Chapter 12: Secure management and maintenance

This section contains the Learning Check answers for Chapter 12.

Learning Check answers

Question 1: AOS-CX switches starting with the CX 6200 Series and up support OOBM. In-band management is often used in campus networks.

Question 2: Administrator, operator, and auditor are the three default AOS-CX user groups.